Virtual Material Acquisition and Representation for Computer Graphics

Synthesis Lectures on Visual Computing

Editor
Brian Barsky, *University of California, Berkeley*

This series presents lectures on research and development in visual computing for an audience of professional developers, researchers, and advanced students. Topics of interest include computational photography, animation, visualization, special effects, game design, image techniques, computational geometry, modeling, rendering, and others of interest to the visual computing system developer or researcher.

Geometric and Discrete Path Planning for Interactive Virtual Worlds
Marcelo Kallmann and Mubbasir Kapadia
2016

An Introduction to Verification of Visualization Techniques
Tiago Etiene, Robert M. Kirby, and Cláudio T. Silva
2015

Virtual Crowds: Steps Toward Behavioral Realism
Mubbasir Kapadia, Nuria Pelechano, Jan Allbeck, and Norm Badler
2015

Finite Element Method Simulation of 3D Deformable Solids
Eftychios Sifakis and Jernej Barbič
2015

Efficient Quadrature Rules for Illumination Integrals: From Quasi Monte Carlo to Bayesian Monte Carlo
Ricardo Marques, Christian Bouville, Luís Paulo Santos, and Kadi Bouatouch
2015

Numerical Methods for Linear Complementarity Problems in Physics-Based Animation
Sarah Niebe and Kenny Erleben
2015

Mathematical Basics of Motion and Deformation in Computer Graphics
Ken Anjyo and Hiroyuki Ochiai
2014

Mathematical Tools for Shape Analysis and Description
Silvia Biasotti, Bianca Falcidieno, Daniela Giorgi, and Michela Spagnuolo
2014

Information Theory Tools for Image Processing
Miquel Feixas, Anton Bardera, Jaume Rigau, Qing Xu, and Mateu Sbert
2014

Gazing at Games: An Introduction to Eye Tracking Control
Veronica Sundstedt
2012

Rethinking Quaternions
Ron Goldman
2010

Virtual Material Acquisition and Representation for Computer Graphics
Dar'ya Guarnera and Giuseppe Claudio Guarnera

ISBN: 978-3-031-01467-3   paperback
ISBN: 978-3-031-02595-2   ebook
ISBN: 978-3-031-00346-2   hardcover

DOI: 10.1007/978-3-031-02595-2

A Publication in the Springer series
A Publication in the Morgan & Claypool Publishers series
Synthesis Lectures on Visual Computing: Computer Graphics, Animation,
Computational Photography, and Imaging

Lecture #30
Series Editor: Brian Barsky, *University of California, Berkeley*
Series ISSN
Print 2469-4215   Electronic 2469-4223

Virtual Material Acquisition and Representation for Computer Graphics

Dar'ya Guarnera and Giuseppe Claudio Guarnera
Norwegian University of Science and Technology

SYNTHESIS LECTURES ON VISUAL COMPUTING #30

ABSTRACT

This book provides beginners in computer graphics and related fields a guide to the concepts, models, and technologies for realistic rendering of material appearance. It provides a complete and thorough overview of reflectance models and acquisition setups, along with providing a selection of the available tools to explore, visualize, and render the reflectance data. Reflectance models are under continuous development, since there is still no straightforward solution for general material representations. Every reflectance model is specific to a class of materials. Hence, each has strengths and weaknesses, which the book highlights in order to help the reader choose the most suitable model for any purpose. The overview of the acquisition setups will provide guidance to a reader who needs to acquire virtual materials and will help them to understand which measurement setup can be useful for a particular purpose, while taking into account the performance and the expected cost derived from the required components. The book also describes several recent open source software solutions, useful for visualizing and manipulating a wide variety of reflectance models and data.

KEYWORDS

virtual materials, reflectance models, reflectance acquisition, BRDF

Contents

Preface

The quest for photorealism drives research toward new developments in Computer Graphics and related fields, with the aim of finally conquering the uncanny valley and making the virtual world indistinguishable from the real one. Achieving a photorealistic virtual representation of the real-world involves several factors, among which is a faithful reproduction of material appearance. Human perception of the appearance of an object depends on the way its surface reflects the incident light, since the reflection provides important visual cues about material properties, such as glossiness and color, used to identify the type of material (plastic, metal, rubber, ceramic, fabric, etc.) and additional information, for instance dry/wet or smooth/rough textures. Visual properties are related to physical characteristics like the diffuse and specular albedo, the index of refraction of the material (as well as that of the physical media in which it is placed, such as air or water), its surface roughness, and so on. The interaction between an opaque, homogeneous surface and the lighting incident on it can be described by the radiometric Bidirectional Reflection Distribution Function (BRDF). A huge body of computer graphics research has been carried out in order to derive compact and accurate models for BRDFs.

Selecting a suitable reflectance model to render a virtual material is not a straightforward task, since each reflectance model often aims to represent specific (subsets of) properties, with the result that a given model can describe plastic very well but not metals, and so on. Not all the BRDF models have become widespread in software packages, and the most common ones can be implemented differently in different renderers, leading to a slightly different rendered image. Hence, in this book we will focus on the theory rather than in the implementation. The reader will find accurate information about BRDF models, their taxonomy and characteristics, and the setup used to acquire the BRDF data, which range from low-to-high cost and low-to-high accuracy.

Dar'ya Guarnera and Giuseppe Claudio Guarnera
December 2017

Acknowledgments

This book is the product of several years of research on Virtual Material Appearance in Computer Graphics, ranging from BRDF acquisition to representation.

We would like to acknowledge people who contributed to our research. First, we would like to express our gratitude to Dr. Mashhuda Glencross, who encouraged and supported us. We are grateful to Dr. Abhijeet Ghosh for the valuable suggestions on current BRDF models and acquisition setups.

A heartfelt thank you to our parents, Alfio, Maria, Alexandr, and Nelli, and our sister Rosalinda and brother Sergey for their love, support, and encouragement. Last but not least a special thanks goes to Anastasiya and Ashanti for all the "input" in the development of this book.

Dar'ya Guarnera and Giuseppe Claudio Guarnera
December 2017

CHAPTER 1

Introduction

The visual appearance of an object is a complex phenomenon, and to describe it properly it is important to understand how a material interacts with the light. For this purpose, many reflectance functions have been investigated, not only in computer graphics, where it is an active research field aiming to obtain photorealistic renderings, but many other disciplines like physics, optical engineering, computer science, and psychology, where considerable time and resources are being invested in the acquisition and representation of reflectance functions. In fact, material appearance plays an important role in many areas of science and industry:

- Computer Vision (e.g., in object recognition applications)

- Aerospace (e.g., for optimal definition of satellite mirrors reflectance and scattering properties)

- Optical Engineering, Remote-Sensing (e.g., land cover classification, correction of view and illumination angle effects, cloud detection, and atmospheric correction)

- Medical Applications (e.g., diagnostics)

- Art (e.g., 3D printing)

- Applied Spectroscopy (e.g., physical condition of a surface)

- Film, Games, Virtual Reality, Marketing, etc.

The world of technology is evolving and offering new, innovative technology that allows the "synthetic world" to look more and more realistic: as a consequence, digital reproduction of real-world material is growing rapidly. These results are driven by the combined effort of researchers, engineers and artists. However, despite the number of material reproduction techniques, material modeling is still a popular topic in research and industry since there is no straightforward way to digitize materials, and often the acquired data is not consistently reusable.

As mentioned, a challenge in computer graphics is how to handle visual appearance accurately measuring and representing material characteristics from the real-world in order to replicate a material behavior and its interaction with the light. Several models and acquisition techniques suggested by researchers in recent years are often aimed at a particular subset of material properties. Ultimately, this limits the applicability of a method only to a specific class of materials. In fact, many methods and setups can properly deal with only a few classes of materials (e.g., leather, fabric, car paint, wood, plastic, rubber, mirrored surfaces, etc.) and unfortunately there is no universal way to acquire and represent all classes. If we look at the material representation side, it could be challenging to choose which a model. On the other hand, digital artists have been provided with applications that include material models with intuitive control, but generally they offer only few material models clearly identifiable with a known analytic formula in the scientific literature. Wider choice of material models is also available in physically based renderers, but those might be difficult to use and far from intuitive. An additional challenge is that the same material rendered in different applications might appear differently due to light tracing algorithms and other implementation choices. The cost of the material digitization could be high or low, where a high cost is not necessarily better. On the material acquisition side, time and costs often represent additional constraints, since they vary enormously across the available measurement devices, and high acquisition time and cost do not necessarily mean "better" (or worse, for that matter).

1.1 CHALLENGES

A major challenge in computer graphics is how to simply and accurately measure the appearance of material characteristics from real-world objects and implement practical editable synthetic materials accurately matching the appearance of the original. Currently, no up-to-date universal material model that can represent leather, fabric, car paint, wood, plastic, rubber, mirrored surfaces, etc. exists [SDSG13]. A variety of rendering algorithms are used in the software pipeline, resulting in a need for optimized material representations, which requires both a flexible acquisition process and representation methods. Unfortunately, the following challenges still persist:

- there is no widely adopted solution;

- few solutions acquire material models that are good enough for a wide range of commercial applications without significant lor and money;

- there is no standardized material model formats from acquisition setups;

- there is little standardization across renderers, with different renderers supporting subsets of material properties;

- material models are hard to edit by artists;

- acquired material models have a high memory footprint which limits applicability;

- it is not easy to assess the accuracy of a material model or a measurement setup.

1.2   SCOPE OF THE BOOK

This book provides an overview of reflection functions, the state-of-the-art material models and reflection acquisition setups, and could be seen as a beginner's guide providing existing techniques for material representation and acquisition in computer graphics, aimed to help in selecting the most appropriate reflectance models and measurement technique for a specific problem. The focus is mainly on current Bidirectional Reflectance Distribution Function (BRDF) representations and acquisition setups, although different reflectance functions will be described where appropriate. For better understanding of the reflectance topic, we begin with a brief overview of the existing reflectance functions, provided in Chapter 2.

Selecting a suitable reflectance model to render a virtual material is not a straightforward task since each reflectance model often aims to represent a specific (subset of) properties, with the result that a given model can describe plastic very well but not metals and so on; BRDF models are described in Chapter 3, where we also describe some tools for BRDF visualization and fitting. In this book we also present different acquisition setups, ranging from low to high cost, each with a varying degree of accuracy. Reflectance acquisition setups are described in Chapter 4, where we guide the reader through available techniques, providing pros and cons of each setup.

For relevant background in mathematics we suggest some additional readings, such as Ström et al. [SAA15]; as for the physics background the reader can find more details in [Gla94].

CHAPTER 2

Reflectance Functions

Human perception of a material depends on how the light that reflected, transmitted, absorbed by an object and reaches the viewer [DR05]. Hence, the appearance of materials may vary significantly depending on a wide range of properties such as color, smoothness, geometry, roughness, reflectance and angle from which the material is viewed and lighting directions.

Clearly not all materials interact with light in the same way: some let part of the light go through the surface, even to the extent of being transparent or semi-transparent; some scatter the light back, toward the light source itself; others show a mirror-like behavior. A ray of light can hit a surface at particular point of an object surface and possibly "travel" under its surface in different directions before leaving the surface in a different spot, after some time, with a totally different direction than initially. The most general reflectance function hence would need to take into account a number of variables (namely 16), which includes the wavelength of the incident ray of light (1 variable) and its direction (2 variables), the time when the ray of light hits the surface (1 variable) and the location on the surface (3 variables), the wavelength (1 variable) and direction (2 variables) of the ray of light leaving the surface at a (possibly) different location on the surface (3 variables) after some time (1 variable), and it must also account for the transmittance direction though the surface (2 variables). These parameters and the parametrization of such a general reflectance function RF are summarized in the following Table 2.1 and Equation 2.1:

$$RF_\lambda \left(\lambda_i, \theta_i, \phi_i, x_i, y_i, z_i, t_i, \lambda_r, \theta_r, \phi_r, x_r, y_r, z_r, t_r, \theta_t, \phi_t, \right). \tag{2.1}$$

The reflectance function RF needs to be measured for each wavelength in the visible spectrum or, at the very least, for the color channels of the RGB color space. Such a reflectance function, schematized in Figure 2.1, can fully describe a material appearance; however, due to its high dimensionality it is currently unfeasible to measure and would produce a vast amount of data, which cannot be handled by current computer graphics and virtual reality applications, considering also that the most general definition of a RF includes the dependence on the polarization state of the incident and reflected light.

For these reasons, in computer graphics and related fields it is customary to rely on several classes of simplified reflectance functions, obtained by discarding some dimensions, more suited for a practical use; in Figure 2.2 we report the taxonomy of the reflectance functions introduced in this section, along with their parameterizations.

The aforementioned simplifications are obtained by assuming the radiance to be constant along the rays of light, which allows discarding the z coordinate of the points on the surface un-

Table 2.1: Parameters of a general reflectance function

Description	Variable(s)
Wavelength the incident ray of light.	λ_i
Direction of the incident ray of light, in spherical coordinates (elevation and azimuth).	$v_i \equiv (\theta_i, \phi_i)$
Time when the ray of light hits the surface.	t_i
3D coordinates of the point in which the ray of light is incident.	$p_i \equiv (x_i, y_i, z_i)$
Wavelength of the outgoing ray of light.	λ_o
Direction of the outgoing ray of light, in spherical coordinates (elevation and azimuth).	$v_o \equiv (\theta_r, \phi_r)$
Time when the ray of light leaves the surface.	t_r
3D coordinates of the point where the ray of light leaves the surface.	$p_r \equiv (x_r, y_r, z_r)$
Transmittance angles (elevation and azimuth).	$\omega_t \equiv (\theta_t, \phi_t)$

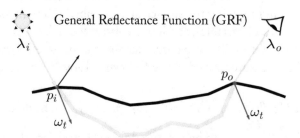

Figure 2.1: Schematic representation of the general reflectance function RF.

der consideration. By dropping the dependency on the time (t_i and t_o, hence assuming that the light transport does not take a measurable time), on the wavelength (λ_i and λ_o), thus assuming that the interaction with the surface does not change the wavelength of the light and restricting our attention to the RGB color bands) and assuming no transmittance ($\theta_t = \phi_t = 0$), we obtain the *Bidirectional Surface Scattering Reflectance Distribution Function* (BSSRDF), the most complex reported in Figure 2.2, which has 8 dimensions. The BSSRDF is able to represent a ray of light incident at a point on the surface, traveling under the surface where it gets scattered in different directions before leaving the surface from a different point and in a different direction. Many common translucent materials like milk, skin and alabaster are characterized by

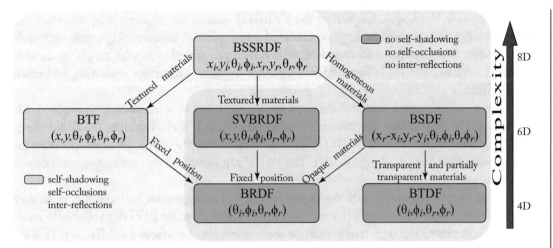

Figure 2.2: **Reflectance functions.**

their subsurface scattering behavior that smooth the appearance of surface details, with the light shining through them. Thanks to its properties the BSSRDF is able to describe phenomena like translucency, self shadowing, self occlusions and inter-reflections. Unfortunately, it is still a very complex function to measure and often simpler representations are preferred over it.

If we assume that the ray o light leaves the surface at the same location where it was incident (hence $x_i = x_r = x$, $y_i = y_r = y$), we obtain the *Bidirectional Texture Function* (BTF), a 6-dimensional representation able to describe not only the local variations in reflectance but also the mesoscopic effects due to small-scale geometry, like self-shadowing, self-occlusions and inter-reflections: $BTF(x, y, \theta_i, \phi_i, \theta_r, \phi_r)$. The term BTF was first introduced by Dana et al. [DVGNK99] as an image-based representation that can describe the fine-scale appearance of a rough surface. The mesoscopic effects are difficult to quantify and separate from the measured data, hence BTF acquisition generally needs a large number of samples of the surface as well as high-end hardware support due to lengthy acquisition times and storage demands [HF13]. Nevertheless, there exist low cost acquisition setups, like the kaleidoscopic device by Han and Perlin [HP03] or the more recent mechanical gantry with rotating arms by Filip et al. [FVK14], built using a toy construction set. BTFs generally result in very realistic material appearance. The first BTF database, described in [DVGNK99], contains 61 real-world surfaces, each observed under 205 different combinations of lighting and viewing illuminations (plus 205 additional measurements for anisotropic surfaces), consisting of over 14.000 images.

A similar parameterizations is used to represent the *Spatially Varying Bidirectional Reflectance Distribution Function* (SVBRDF), used to describe opaque, smooth materials that can have different reflectance at each point of the surface (non-homogeneous materials) [HF13]. The SVBRDF parameterization hence must takes into account the location over the surface:

$SVBRDF(x, y, \theta_i, \phi_i, \theta_r, \phi_r)$. Capturing the SVBRDF sometimes requires long measurements and processing times as well as large, specialized and expensive hardware rigs, although under certain assumptions approximate measurements can be performed even with cellphone or tablet cameras [AWL15, RPG15]. The SVBRDF cannot describe subsurface scattering and mesoscopic effects.

For a homogeneous material that can reflect light but also transmit it through its surface we need to reintroduce the transmittance angles (θ_t, ϕ_t), thus obtaining the Bidirectional Scattering Distribution Function (BSDF), comprising scattering effects for both reflection and transmission: $BSDF(\theta_i, \phi_i, \theta_r, \phi_r, \theta_t, \phi_t)$. The BSDF can describe both transparent and opaque materials.

If we take into account only the transmittance of homogeneous material it is possible to describe it with the *Bidirectional Transmittance Distribution Function* (BTDF), suitable to model how the light passes through transparent or semi-transparent surfaces [WMLT07], [HF13]: $BTDF(\theta_i, \phi_i, \theta_t, \phi_t)$.

An opaque, smooth, homogeneous material can be represented with the *Bidirectional Reflectance Distribution Function* (BRDF): $BRDF(\theta_i, \phi_i, \theta_r, \phi_r)$. By looking at Figure 2.2 it is possible to note how the BRDF can be considered a special case of the more complex functions described above [ASMS01]. In fact, the BRDF can be considered as a special case of the BTF and the SVBRDF when the position on the sample surface is fixed; any BTF datasets can be approximated as a sparse linear combination of rotated analytical BRDFs [WDR11] and the SVBRDF parameterization includes extra parameters with respect to the BRDF simply to take into account the location over the surface, but it must fulfill the BRDF reciprocity and energy conservation properties, which will be described in the next sections. Finally, a BSDF can be modeled as a sum of a BRDF (for the reflection component) and a BTDF (for the transmittance component).

2.1   DEFINITION OF THE BRDF

As discussed in the previous section, one of the possible ways to represent the way an opaque, homogeneous material interacts with the light is through the BRDF (Bidirectional Reflectance Distribution Function), a radiometric function currently used to varying levels of accuracy in photorealistic rendering systems. It describes, in the general case, how incident energy redirects in all directions across a hemisphere above the surface. Historically, the BRDF was defined and suggested over the more generalized BSSRDF (Bidirectional Scattering Surface Reflectance Distribution function) [JMLH01] by Nicodemus [NRH*77], as a simplified reflectance representation for opaque surfaces: the BRDF assumes that light entering a material leaves the material at the same position, whereas the BSSRDF can describe light transport between any two incident rays on a surface. Many common translucent materials like milk, skin and alabaster cannot be represented by a BRDF since they are characterized by their subsurface scattering behavior that smooths the surface details, with the light shining through them [GLL*04].

These materials are expensive to measure and render. However many techniques have been proposed [JMLH01], [DS03], [HBV03], [DWd*08], [DI11], [KRP*15].

Before defining the reflectance we provide a brief introduction to some important radiometric terms. *Radiant Energy*, the basic unit of energy, is measured in *Joules* [*J*] and indicated with the symbol Q:

$$Q = \frac{hc}{\lambda} \quad [J],$$ (2.2)

where h is the Planck's constant, c is the speed of light in vacuum and λ is the wavelength of the incident photon. The energy flowing through a surface per unit time is called *Radiant Flux*, indicated with Φ and measured in Watts [*W*]:

$$\Phi = \frac{dQ}{dt} \quad [J/s = W].$$ (2.3)

The flux flowing per unit of surface area is called *radiant flux area density*, measured [W/m^2] and indicated by u:

$$u = \frac{d\Phi}{dA} \quad [W/m^2].$$ (2.4)

Two different terms are used in order to distinguish the flow of energy toward a surface from the flow leaving a surface: in the first case we refer to *irradiance* (E); in the second one the term used is *radiosity* (B). If instead of referring to the ratio of flux per unit of surface area we take into account a solid angle, we can define the intensity I, the radiant energy leaving a point in the direction Φ per unit solid angle, measured in [W/sr]:

$$I = \frac{d\Phi}{d\vec{\omega}} \quad [W/sr].$$ (2.5)

Finally, the radiant flux per unit solid angle and per unit projected area is called *radiance*:

$$L = \frac{d^2\Phi}{d\omega \, dA \cos\theta} \quad \left[\frac{W}{sr \cdot m^2}\right].$$ (2.6)

In the following we indicate the radiance arriving at a surface with L_i and the radiance leaving a surface with L_r.

We are now ready to define the BRDF as the ratio of the reflected radiance L_r to incident irradiance E:

$$f_r(\mathbf{v_i}, \mathbf{v_r}) = \frac{dL_r(\mathbf{v_r})}{dE_i(\mathbf{v_i})},$$ (2.7)

where $\mathbf{v_i}$ and $\mathbf{v_r}$ are vectors describing the incident (i) and exitant (r) directions. By taking into account the incident radiance L_i instead of E_i, thus considering the solid angle around the incident lighting direction and the cosine of the angle between the latter and the surface normal, we can write the Equation (2.7) in a different form, which allows understanding how the units of a BRDF are inverse steradian [$1/sr$]:

$$f_r\left(\mathbf{v_i}, \mathbf{v_r}\right) = \frac{dL_r\left(\mathbf{v_r}\right)}{L_i\left(\mathbf{v_i}\right)\cos\theta_i\, d\omega_i}. \tag{2.8}$$

Researchers have measured hundreds of BRDFs, suggested implementation techniques and utilized user input to edit and enhance materials. Recent implementations have expanded material libraries but have not improved significantly upon material representation efficiency. However, the uptake of acquired models has not been widespread across rendering packages due to their data and storage requirements.

To understand the way the BRDF is parameterized, let's take into consideration a point p on a surface and the surface normal $\mathbf{n}$ at that specific location on the surface; on the plane tangent to the surface in p we fix a reference direction $\mathbf{t}$, called tangent direction, and its perpendicular direction $\mathbf{b}$ on the plane: $\mathbf{n} \times \mathbf{t} \times \mathbf{b}$ defines a local reference frame. Once we set the incoming light direction and the outgoing direction (viewing direction), the angle between the surface normal and the viewing direction is called θ_i; similarly the angle between the surface normal and the outgoing direction is called θ_r. If we take the projection of the viewing direction on the tangent plane, the angles between the tangent direction and the projection of the incoming direction are called respectively ϕ_i, and ϕ_r.

Figure 2.3 shows the geometry of the BRDF and the vectors used for parameterizations:

- $\mathbf{n}$ is the normal at a specific point p on the surface.

- $\mathbf{t}$ is the tangent vector. It is perpendicular to the normal $\mathbf{n}$ and hence it is tangent to the surface at p.

- $\mathbf{b}$ is the bi-tangent vector, defined as $\mathbf{b} = \mathbf{n} \times \mathbf{t}$. In literature it is also named as binormal vector.

- $\mathbf{h}$ is the halfway vector [Rus98], defined as: $\mathbf{h} = \frac{(\mathbf{v_i}+\mathbf{v_r})}{\|\mathbf{v_i}+\mathbf{v_r}\|}$.

Another very common way to parameterize the BRDF is the halfway $\mathbf{h}$ vector shown in Figure 2.4, defined by the normalized vector sum of the incoming and outgoing directions. In this case we are taking into account the angle between the surface normal $\mathbf{n}$ and the halfway vector $\mathbf{h}$. This has important implications in the way the measured data can be stored, compressed and can speed up computation of specific models. The use of the halfway vector enables another possibility to define a local reference frame, in which one of the axes is aligned with $\mathbf{h}$ and the other two are given by $\mathbf{b'} = \frac{(\mathbf{n} \times \mathbf{h})}{\|\mathbf{n} \times \mathbf{h}\|}$ and $\mathbf{t'} = \mathbf{b'} \times \mathbf{h}$.

There exist other coordinate systems and parameterizations especially suited for dimensionality reduction of some isotropic BRDF models, for instance the barycentric coordinate system with respect to a triangular support proposed by Stark et al. [SAS05], or the hybrid model described by Barla et al. which could lead to a better repartition of samples to cover most of the effects of materials [BBP15].

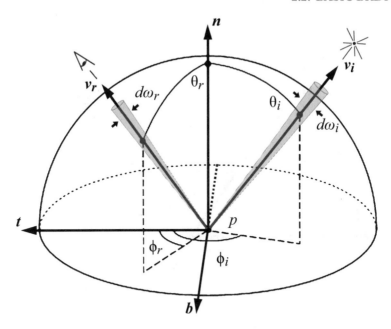

Figure 2.3: Geometry of BRDF.

2.2   BASIC BRDF MODELS

There are many reflectance models that are simplified subsets of the BRDF function. One of the simplest reflectance models is the Lambertian model, which represents the perfect diffuse reflectance and is often used in many interactive applications, since it requires no recalculation with the change of viewing direction. The model simply assumes that the surface reflects light uniformly in all directions with the same radiance (see Figure 2.5, in yellow), constant with v_r, unlike other BRDF models: $f_r(\mathbf{v_i}, \mathbf{v_r}) = \rho_d/\pi$, where ρ_d is the diffuse albedo.

In the case of a pure specular BRDF all the light is reflected in a single direction for a given incident direction (see Figure 2.5, in light blue). In fact, light that is incident within a differential solid angle $d\omega_i$ from direction (θ_i, ϕ_i) is reflected in a differential solid angle ω_r in direction $(\theta_i, \phi_i + \pi)$, hence the pure specular BRDF can be formalized with a double Dirac delta function: $f_r(\mathbf{v_i}, \mathbf{v_r}) = \rho_s \delta(\theta_i - \theta_r)\delta(\phi_i + \pi - \phi_r)$, where $\rho_s = L_r/L_i$ is the specular albedo. Perfect specularity is valid only for highly polished mirrors and metals.

Surfaces not perfectly smooth, which have some roughness at the micro-geometry level, have a glossy appearance and show broader highlights, other than specular reflections (see Figure 2.5).

Some materials, like the surface of the moon or some biological tissues, show a phenomenon called retro-reflection in which light is scattered not only in the forward direction but

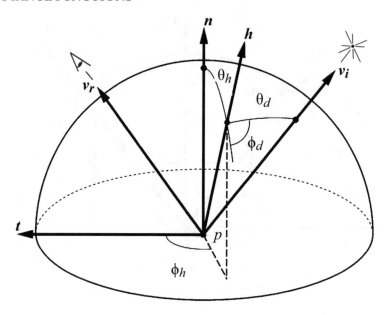

Figure 2.4: Halfway vector parameterization.

also in the direction of the illuminant. Real material tends to display a mixture of the aforementioned basic reflectance types, giving rise to very complex reflection properties.

BRDFs can be classified by taking into account the characteristics of the reflection, whether they change by rotating the surface around its normal direction:

- *Isotropic* BRDFs are able to represent materials whose reflection does not depend on the orientation of the surface, since the reflectance properties are invariant to rotations of the surface around **n**.

- *Anisotropic* BRDFs can describe materials whose reflection changes with respect to rotation of the surface around **n**; this class includes materials like brushed metal, satin, velvet and hair.

The Fresnel effect predicts the fraction of power that is reflected and transmitted and has a great impact on the appearance (Figure 2.7). Many basic BRDF models have lost importance in the context of physically based modeling because they do not account for a Fresnel term. For conductive materials, like metals, the fraction of light reflected by pure specular reflection is roughly constant for all angles of incidence, whereas for non-conductive materials (dielectrics), the amount of light reflected increases at grazing angles; see Figure 2.6 for a comparative example of the behavior of metals and dielectrics. The fraction of light reflected is called Fresnel reflectance, which can be obtained from the solution of Maxwell's equations and depends also on the polarization state of the incident light. For unpolarized light, the Fresnel reflectance F

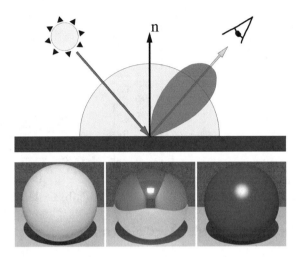

Figure 2.5: Basic reflectance models of the incoming light (in orange): perfect diffuse (yellow), glossy (purple) and perfect specular (light blue). Renderings of diffuse, glossy and specular spheres are shown, placed inside a Cornell box [GTGB84].

at the interface between the surface and the air is given by

$$F(\eta, \theta_i, \theta_t) = \frac{1}{2}\left[\left(\frac{\eta\cos\theta_i - \cos\theta_t}{\eta\cos\theta_i + \cos\theta_t}\right)^2 + \left(\frac{\cos\theta_i - \eta\cos\theta_t}{\cos\theta_i + \cos\eta\theta_t}\right)^2\right], \qquad (2.9)$$

where η is the index of refraction of the surface and θ_t is the angle of transmission. In computer graphics, it is very common to use Schlick's approximation of the Fresnel reflectance [Sch94]:

$$F(\theta) = F(0) + (1 - F(0))(1 - \cos(\theta))^5, \qquad (2.10)$$

where $F(0)$ is the Fresnel reflectance at normal incidence; in the following chapters we will generally use the symbol F to refer either to the exact Fresnel reflectance or one of its approximations.

A BRDF should respect some basic physical properties, namely non-negativity, reciprocity and energy conservation:

- non-negativity: the BRDF is a non-negative function, hence for any pair of incident and outgoing direction $f_r(\mathbf{v_r}, \mathbf{v_i}) \geq 0$;

- the Helmholtz reciprocity principle states that the light path is reversible for any pair of incident and outgoing direction: $f_r(\mathbf{v_i}, \mathbf{v_r}) = f_r(\mathbf{v_r}, \mathbf{v_i})$. This principle holds only for corresponding states of polarization for incident and emerging fluxes, whereas large discrepancies might occur for non-corresponding states of polarization [CP85]. In designing a rendering system possible non-reciprocity should be taken into account [Vea97].

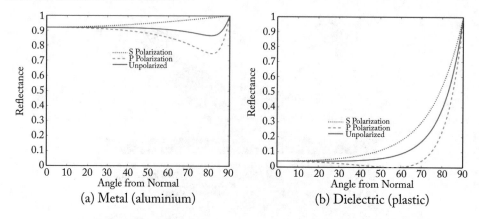

Figure 2.6: Fresnel reflectance for metals (a) and dielectrics (b).

(a) Without Fresnel   (b) With Fresnel

Figure 2.7: A dielectric sphere rendered without accounting for the Fresnel reflectance (a) and accounting for it (b).

- Energy conservation assumes that the energy reflected cannot exceed incident energy [DRS07]: $L_r \leq E_i$, hence over the unit hemisphere Ω_+ above the surface

$$\forall \mathbf{v_i}, \int_{\Omega_+} f_r(\mathbf{v_i}, \mathbf{v_r})(\mathbf{v_r} \cdot \mathbf{n}) d\omega_r \leq 1. \tag{2.11}$$

CHAPTER 3

Models of BRDF

In the last 40 years many material models have been proposed and some BRDFs are able to describe a wide subset of a material properties we have mentioned. Generally they can be classified into three big families that have different purposes and different ways of calculating interaction with light:

- Phenomenological

- Physically based

- Data driven

Phenomenological models, also known as Empirical models, are entirely based on reflectance data, which is fitted to analytical formulas, thus approximating the reflectance and reproducing characteristics of real world materials, but they will not necessarily appear realistic unless placed in an accurately simulated environment for such a model.

Physically based models are based on Physics and Optics, with the assumption that the surface is rough at a fine scale, therefore described by a collection of micro facets with some distribution D of size and direction; these models are also referred to as "first principle" models. Usually they are represented by accurate and adjustable formulae; however the most common mathematical model has the form:

$$f_r(\mathbf{v_i}, \mathbf{v_r}) = \frac{D \cdot G \cdot F}{4 \cos \theta_i \theta_r} \tag{3.1}$$

which also takes into account the Fresnel term F. Effects like masking and self-shadowing (see Figure 3.1) [AMHH08] depend on the projected area of the microfacets and hence on the distribution D, generally described by the geometrical attenuation term G; for a review of common masking functions and a derivation of the exact form of the masking function from the microsurface profile, see the work by Heitz [Hei14]. This class of models can represent unique properties of the material and may include subsurface structure, generally resulting in complex calculations due to the interaction of the light with the surface structure.

Data-Driven models approximates measured BRDFs with a suitable function space, for example, spherical harmonics or wavelets, weighted sum of separable functions or product of functions. Measured BRDF data, produced by most of the setups described in Chapter 4, can be stored in a table or a grid and then interpolated to produce a large look-up table when data is needed. This method is simple but inefficient in terms of storage. Moreover, the measured

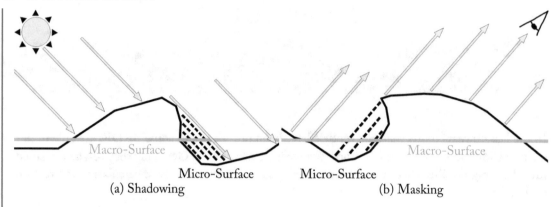

Figure 3.1: (a) Due to the microgeometry, some microfacets are occluded and do not receive light (shadowing). (b) The light reflected from microfacets not visible from the viewing direction cannot be seen (masking).

raw data is often noisy, hence the noise is likely to appear in the rendered material. A measured BRDF can be fitted to analytic models and employed to reconstruct the BRDF, thus significantly reducing storage size. The downside of this strategy is related to the inflexibility of many models, which are hard to edit and able to represent only limited classes of materials. A different solution is to approximate measured BRDFs with a suitable function space, for example, spherical harmonics or wavelets, weighted sum of separable functions, or product of functions. We refer to this class of models as Data-Driven models, described in Section 3.3.

Many media, such as hair, fur, cloth, and knitwear are difficult to describe by a surface model. These materials, and objects with highly complex boundaries, are better described by volumetric appearance models [KK89, PH89, XCL*01], in particular for closer viewing distance, whereas BRDFs can be used from farther away, as demonstrated by Guarnera et al. [GHCG17] in their technique for fast reverse engineering of woven fabrics at a yarn level, suitable for rapid acquisition of woven cloth structure and making use of a surface-based representation. Jakob et al. [JAM*10] introduced a generalization to anisotropic scattering structures, exploited also for volumes acquired by CT scans [ZJMB11]. More recently, collections of individual fibers have been used for fabric representation [KSZ*15]. In this book we focus on representation and acquisition of surface reflectance, hence we do not further discuss volumetric representation.

A very important aspect is the practicality of a model in a rendering system, which requires a suitable technique for importance sampling. When calculating the radiance direction of a surface in a scene, accounting for the contribution of light from all possible directions is expensive to compute. Therefore Monte Carlo techniques are used to estimate the values with fewer samples [Hai91], based on a stochastic process. However, the number of samples should be sufficient to produce consistent estimations; otherwise the results will vary significantly. Importance sampling can be used to reduce sample variance [LRR04] by distributing samples according to the known elements, either taking into account the reflection model in use or the incident light [CJAMJ05].

Tables 3.1, 3.2, and 3.3 summarize the characteristics of the reflectance models detailed in the following sections, selected to widely cover as many possible unique models proposed in the computer graphics and Vision literature, suitable for a broad range of materials. Figure 3.2 reports the taxonomy of reflectance models, depicted more in detail in Figures 3.3, 3.8, and 3.15.

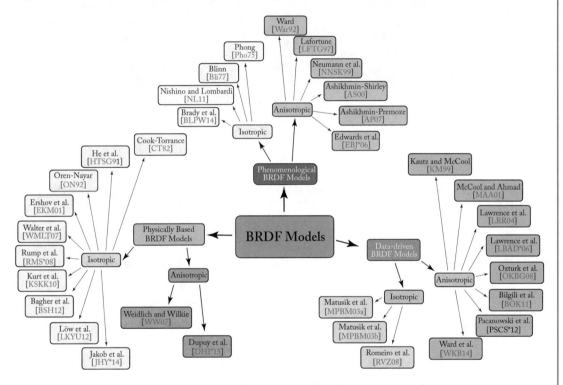

Figure 3.2: Taxonomy of reflectance models color coded for isotropic and anisotropic representations.

Table 3.1: Some of the main properties of the BRDF models described in Section 3.1. Models are grouped by category (isotropic/anisotropic) and sorted by year.

	Model and Ref	Reciprocity	Energy Cons.	Fresnel	Sampling	Short Description
	Phenomenological					
Isotropic	Phong [Pho75]	✗	✗	✗	✓	Basic surface representation model based on the Cosine law.
	Blinn–Phong [Bli77]	✗	✗	✗	✓	Based on [Pho75], uses the halfway reflection direction for faster computation.
	Nishino and Lombardi [NL11]	✓	✓	✗	✗	Models BRDFs as a mixture of hemispherical distribution functions; small footprint.
	Brady et al. [BLPW14]	✓	✗	✓	✗	Framework for automatic learning of analytical models. Some of the properties are not guaranteed by the grammar.
Anisotropic	Ward [War92]	✓	✗	✗	✓	Versatile and cheap to compute.
	Lafortune [LFTG97]	✓	✓	✗	✓	Generalization of the cosine lobe model with multiple steerable lobes.
	Neumann et al. [NNSK99]	✓	✓	✓	✓	Physically plausible formulation of the Phong, Blinn-Phong, and Ward models.
	Ashikhmin–Shirley [AS00]	✗	✗	✓	✓	Based on [Pho75], includes anisotropic reflections for two-layered materials.
	Edwards et al. [EBJ*06]	✗	✓	✗	✓	Framework for transforming the halfway vector into different domains.
	Ashikhmin-Premoze [AP07]	✓	✓	✓	✓	Combines analytic model with a data-driven distribution; accounts for backscattering.

Table 3.2: Some of the main properties of the physically based BRDF models described in Section 3.2. Models are grouped by category (isotropic/anisotropic) and sorted by year.

	Model and Ref	Reciprocity	Energy Cons.	Fresnel	Sampling	Short Description
	Physically Based					
Isotropic	Cook-Torrance [CT82]	✓	✗	✓	✗	Can model metals and plastics, view dependent changes.
	He et al. [HTSG91]	✓	✓	✓	✗	Enhances [CT82], allowing more general material representation.
	Oren-Nayar [ON94]	✓	✓	✗	✓	Enhances the Lambertian model for rough diffuse surfaces.
	Ershov et al. [EKM01]	✗	✓	✓	✗	Focuses on layered materials, like metallic paint. It models binder pigment particles, flakes, and flake coating.
	Walter et al. [WMLT07]	✓	✓	✓	✓	Defines the GGX distribution; based on a BSDF representation.
	Rump et al. [RMS*08]	✗	✗	✓	✓	Suitable for metallic paints, combines [CT82] for the base layer with BTF for top paint layer, including particles.
	Kurt et al. [KSKK10]	✓	✓	✓	✓	Multiple specular lobe model can represent layered or mixed materials.
	Bagher et al. [BSH12]	✓	✗	✓	✓	Provides accurate fitting for materials in the MERL database.
	Löw et al. [LKYU12]	✓	✗	✓	✓	Guarantees accurate fitting to measured data for glossy surfaces; describes 2 models based on the ABC distribution.
	Jakob et al. [JHY*14]	✗	✗	✓	✓	Allows modeling spatially varying BRDF appearance of glittery surfaces.
Anisotropic	Weidlich and Wilkie [WW07]	✓	✓	✓	✓	Multi-layered model that includes absorption and internal reflection.
	Dupuy et al. [DHI*15]	✗	✗	✓	✓	Method to automatically convert a material to a microfacet BRDF.

Table 3.3: Some of the main properties of the data-driven BRDF models described in Section 3.3. Models are grouped by category (isotropic/anisotropic) and sorted by year.

	Model and Ref	Reciprocity	Energy Cons.	Fresnel	Sampling	Short Description
				Data-Driven		
Isotropic	Matusik et al. [MPBM03a]	✓	✓	✗	✓	Provides realistic appearance and meaningful parameterization.
	Matusik et al. [MPBM03b]	✓	✗	✗	✓	Reduces number of samples to acquire and represent BRDF.
	Romeiro et al. [RVZ08]	✓	✗	✓	✓	Bivariate representation, allows to capture off-specular and retro-reflections.
Anisotropic	Kautz and McCool [KM99]	✗	✗	✗	✓	SVD- or ND-based decomposition for BRDFs; approximation based on textures, used to store directions.
	McCool and Ahmad [MAA01]	✓	✓	✗	✓	Based on logarithmic homomorphism. Simple parameterization and limited storage cost (2 textures).
	Lawrence et al. [LRR04]	✗	✗	✗	✓	Provides accurate results and can be also used for BTFs; compact representation.
	Lawrence et al. [LBAD*06]	✓	✓	✓	✗	Suited for interactive rendering/editing.
	Ozturk et al. [OKBG08]	✓	✗	✗	✗	Computationally efficient linear model for approximating BRDFs.
	Bilgili et al. [BÖK11]	✗	✗	✓	✓	Recursive application of the Tucker decomposition on the error term.
	Pacanowski et al. [PSCS*12]	✓	✗	✓	✓	Projects measured BRDFs on a 2D space and approximates them with Rational Functions; small footprint.
	Ward et al. [WKB14]	✓	✗	✗	✓	Tensor tree representation for measured BSDF data.

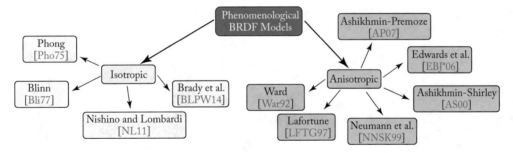

Figure 3.3: Phenomenological isotropic and anisotropic models described in Section 3.1.

3.1   PHENOMENOLOGICAL MODELS

3.1.1   PHENOMENOLOGICAL MODELS FOR ISOTROPIC MATERIALS

One of the simplest BRDF models was proposed by Phong [Pho75] to describe the reflectance of isotropic materials with a slightly rough surface. The specular reflection is modeled by means of a specular constant, the dot product between the viewing direction and the mirror direction of the incoming light, and an exponent n that controls the width of the specular lobe (higher values of n give narrower lobes and a more specular appearance): $f_r(\mathbf{v_i}, \mathbf{v_r}) = k_s(\mathbf{v_r} \cdot \mathbf{r_{vi}})^n$, where k_s is a specular constant in the range $[0, \infty]$, $\mathbf{r_{vi}}$ is the direction of $\mathbf{v_i}$ after being perfectly reflected, and the exponent n controls the shape of the specular highlight. This model does not take into account energy conservation or reciprocity. Moreover, it does not capture the reflection behavior of real surfaces at grazing angles, since it does not account for the Fresnel effect. A set of renderings for increasing values of n are reported in Figure 3.4. Some normalization factors for cosine lobes have been proposed, either based on double-axis moments [Arv95] or with the simpler option of a power series in $(\mathbf{n} \cdot \mathbf{h})$ with a suitable sequence of exponents [Lew94].

Blinn [Bli77] suggested a correction to the Phong model. It is still based on the cosine lobe model. However the dot product is calculated between the surface normal and the halfway vector. In those years it was costly to calculate the mirror direction of the incoming light, which needs to be calculated at each position of the scene; thanks to the modification proposed by Blinn, when an orthographic camera is used, and hence the halfway vector can be approximately considered to be constant across the scene, the computational cost is greatly reduced. Although it has been used as the default shading model for OpenGL and Direct3D until recent times (mostly because of the aforementioned reduction in the computational cost), it shares the same limitations of the Phong model, being physically not plausible: it does not fulfill reciprocity nor energy conservation and it does not account for the Fresnel effect.

$$f_r(\mathbf{v_i}, \mathbf{v_r}) = k_s(\mathbf{n} \cdot \mathbf{h})^n. \tag{3.2}$$

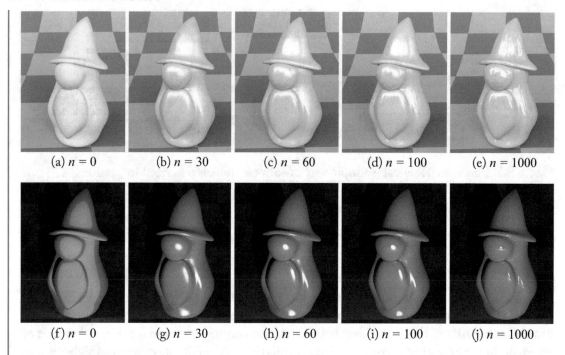

(a) $n = 0$ (b) $n = 30$ (c) $n = 60$ (d) $n = 100$ (e) $n = 1000$

(f) $n = 0$ (g) $n = 30$ (h) $n = 60$ (i) $n = 100$ (j) $n = 1000$

Figure 3.4: Phong model. The top row reports a sequence of renderings for increasing values of n, under environment map lighting; from left to right: $n = 30, 60, 100, 1000$. The bottom row shows the same sequence of exponents, with a point light illumination.

Since it follows the cosine function, if n goes to infinity the reflected radiance and the albedo converges to zero toward grazing angles.

Nishino and Lombardi [NL11] proposed a model in which the surface reflectance is described by multiple lobes, whose optimal number k is automatically computed by an Expectation-Maximization algorithm. A BRDF is treated as a statistical distribution on a unit hemisphere, allowing to model an isotropic BRDF with a mixture of low-dimensional directional statistics distributions over the hemisphere. Such low-dimensional directional statistics distributions are called Hemispherical Exponential Power Distribution (Hemi-EPD), probability density functions that take in input a direction for $\mathbf{v}_i$ and return a distribution of directions for $\mathbf{v}_i$. The Hemi-EPD constitutes a basis for the entire BRDF, which can be modeled as mixtures of Hemi-EPDs, one for each of its 2D slices. The BRDF is parameterized using the halfway vector and difference angle; thanks to the isotropy the dependence on ϕ_h can be dropped. The expression for the Hemi-EPDs, given the parameters $\Theta = \kappa, \gamma$ and the normalization factor C, is $p(\theta_h | \theta_d, \Theta) = C(\Theta)(e^{\kappa \cos^\gamma \theta_h} - 1)$. Energy conservation is guaranteed when the condition $\sum_1^k 1/C(\Theta_k) \leq 1$ holds.

Genetic programming can be used to learn new analytical BRDF models, as demonstrated by Brady et al. [BLPW14]. The training set consists of eight isotropic materials from the MERL dataset [MPBM03a], whereas the starting seeds are a few simple BRDF models. The algorithm applies symbolic transformations to the seeds, and the fitness function calculates the residual error of each variant after fitting the free parameters to the training set. The random search is heuristic-based, trying to adapt the starting models to the measured ground truth data. The algorithm tends to produce a large set of candidate expressions, and to allow a better exploration of the search space, some suboptimal variations that increase the error are allowed, together with an island model genetic algorithm, which sporadically allows interactions between sub-populations. The grammar does not guarantee that the resulting models respect energy conservation and reciprocity, hence these properties need to be taken into account by the fitness function; a table with variants for which energy conservation and reciprocity have been numerically verified is reported in [BLPW14].

3.1.2   PHENOMENOLOGICAL MODELS FOR ANISOTROPIC MATERIALS

The Ward reflectance model [War92] is able to represent both isotropic and anisotropic reflection, and an efficient for Monte Carlo sampling strategy can be derived for the model; it combines specular and diffuse components of reflectance, representing specular peaks through Gaussian distributions. It was specifically designed to easily fit measured BRDFs, which have been used for its validation. The Ward model has four parameters, which can be set independently; therefore it can be fitted to a large class of measured data. The anisotropic model makes use of the two parameters α_x and α_y to control the width of the Gaussian lobe in the two principal directions of anisotropy:

$$f_r\left(\mathbf{v_i}, \mathbf{v_r}\right) = \frac{\rho_d}{\pi} + \frac{\rho_s}{\sqrt{\cos\left(\theta_i\right)\cos\left(\theta_r\right)}} \cdot \frac{e^{-\tan^2\left(\theta_h\right)\left(\frac{\cos^2\theta_h}{\alpha_x{}^2} + \frac{\sin^2\theta_h}{\alpha_y{}^2}\right)}}{4\pi\alpha_x\alpha_y}, \qquad (3.3)$$

where ρ_s controls the magnitude of the lobe and $4\pi\alpha^2$ is a normalization factor; the isotropic version of the model is obtained by setting $\alpha_x = \alpha_y$. In Figure 3.5 we report the shape of the isotropic Ward specular lobe for three different values of surface roughness $\alpha_x = \alpha_y$.

The model does not obey the principle of energy conservation at grazing angles; this behavior has been further investigated in later work such as [NNSK99], [Dür06], [GMD10]. In particular, a different normalization factor has been proposed in [Dür06] to prevent numerical instabilities and to correct the loss of energy at flat angles, specifically $(4\cos\left(\theta_i\right)\cos\left(\theta_r\right))$ instead of $\left(4 \cdot \sqrt{\cos\left(\theta_i\right)\cos\left(\theta_r\right)}\right)$. However it shares the problem of diverging to infinity with the original Ward model. A new physically plausible version of the model has been proposed in [GMD10] (see Figure 3.6), which meets the energy conservation principle even at grazing angles by using the following normalization factor:

$$\frac{2\left(1 + \cos \theta_i \cos \theta_r + \sin \theta_i \sin \theta_r \cos \phi_r - \phi_r\right)}{\left(\cos \theta_i \cos \theta_r\right)^4}. \tag{3.4}$$

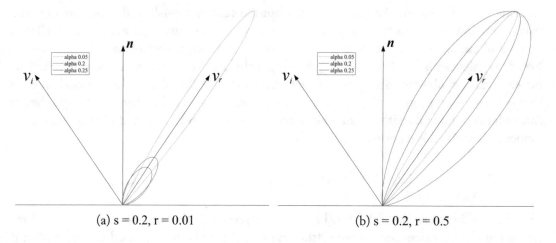

(a) s = 0.2, r = 0.01   (b) s = 0.2, r = 0.5

Figure 3.5: Isotropic Ward specular lobes for three different values of roughness. In (a) the Gaussian distributions corresponding to the lobes have been normalized in order to have the same area; in (b) the lobes have been normalized in order to have the same specular peak.

The Lafortune [LFTG97] model is a flexible, empirical model designed to fit measurements from real surfaces and compactly represent them [WLT04]. The model is a generalization of the cosine lobe model with multiple steerable lobes, based on the Phong model. The primitive functions obey the Energy Conservation and Reciprocity principles. This model allows lobe specification on the surface in terms of shape and direction by simply setting up to three parameters and an exponent:

$$f_r\left(\mathbf{v_i}, \mathbf{v_r}\right) = \frac{\rho_d}{\pi} + \sum_{l=1}^{N} \left(C_{x,l} v_{i\,x} v_{r\,x} + C_{y,l} v_{i\,y} v_{r\,y} + C_{z,l} v_{i\,z} v_{r\,z}\right)^{n_l}, \tag{3.5}$$

where N is the number of lobes, and C_x, C_y, C_z are parameters that absorb the specular albedo and control retro-reflections (by setting C_x, C_y and C_z to positive values), anisotropy (with $C_x \neq C_y$) and off-specular peaks (if C_z is smaller than $-C_x = -C_y$). Lafortune's reflection model can represent generalized diffuse reflectance as the model is able to reflect radiance evenly in all directions by setting $C_x = C_y = 0$; the Lambertian model can be obtained by setting $N = 0$. A comparative study shows that the Lafortune model performs better than the Phong, Ward, and He et al. models in representing measured BRDFs like white paper, rough plastic, rough aluminium, and metal [WLT04].

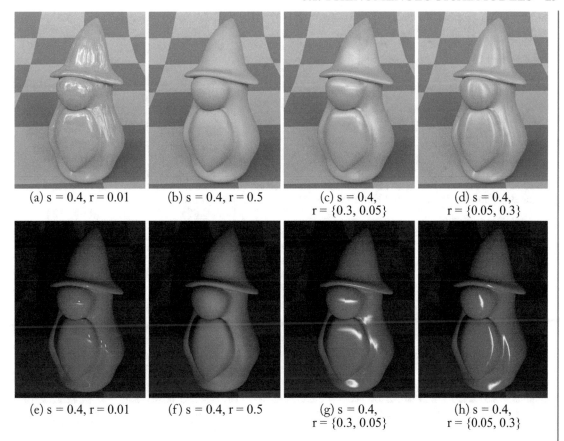

(a) s = 0.4, r = 0.01   (b) s = 0.4, r = 0.5   (c) s = 0.4, r = {0.3, 0.05}   (d) s = 0.4, r = {0.05, 0.3}

(e) s = 0.4, r = 0.01   (f) s = 0.4, r = 0.5   (g) s = 0.4, r = {0.3, 0.05}   (h) s = 0.4, r = {0.05, 0.3}

Figure 3.6: Energy-conserving Ward model variant [GMD10]. The first two columns refer to an isotropic material, whereas the last two columns refer to an anisotropic material. The top row reports a sequence of renderings under environment map lighting. The bottom row shows the same sequence under a point light illumination. The parameters used are reported in brackets: the first value (s) refers to the specular reflectance, the second value (r) to the isotropic roughness, or in case of anisotropic material, the second value refers to the roughness in the tangent direction, the third one to the bitangent direction.

Neumann et al. [NNSK99] proposed some modifications and correction factors for the reciprocal Phong [Pho75], [LW94], Blinn [Bli77] and Ward [War92] models. The correction factors can be seen as shadowing and masking terms to make the models physically plausible. Moreover the modified models can be used to render metals and other specular objects and for each of them an importance sampling procedure is described.

The Ashikhmin-Shirley model (see Figure 3.7) makes use of Phong-like exponents in order to control the shape of the specular lobe and provides a Fresnel-dependent expression for

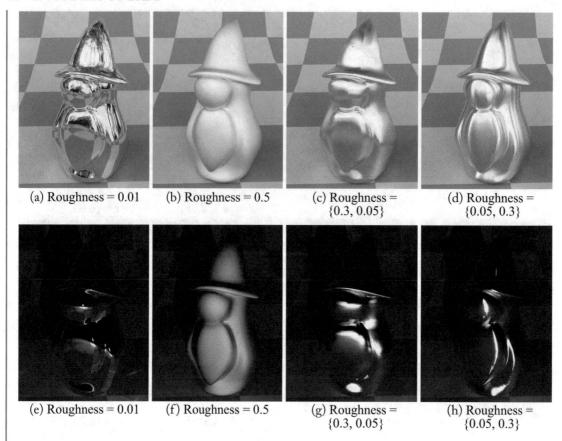

(a) Roughness = 0.01   (b) Roughness = 0.5   (c) Roughness = {0.3, 0.05}   (d) Roughness = {0.05, 0.3}

(e) Roughness = 0.01   (f) Roughness = 0.5   (g) Roughness = {0.3, 0.05}   (h) Roughness = {0.05, 0.3}

Figure 3.7: A set of rendering of gold with the Ashikhmin-Shirley model [AS00]. The first two columns refer to an isotropic surface, whereas the last two columns refer to an anisotropic one. The top row reports a sequence of renderings under environment map lighting; the bottom row shows the same sequence under a point light illumination. The surface roughness values are reported in brackets; in case of anisotropic surface the first value refers to the roughness in the tangent direction, the second one to the bitangent direction.

both diffuse and specular reflections [AS00]; the reflectance of the model changes with the viewpoint, hence at grazing angles the reflectance is specular and at normal angles the reflectance is diffuse. This model assumes microfacets with various angles and sizes [OKBG08] by generalizing the types of microfacets and allows the expression of arbitrary angles.

The specular component of the BRDF is expressed by:

$$f_{r,s}(\mathbf{v_i}, \mathbf{v_r}) = \frac{F(\mathbf{v_i} \cdot \mathbf{h}) \, D(\mathbf{h})}{2(\mathbf{h} \cdot \mathbf{v_r}) \max(\mathbf{n} \cdot \mathbf{v_r}, \mathbf{n} \cdot \mathbf{v_i})}, \qquad (3.6)$$

where $F\,(\mathbf{v_i} \cdot \mathbf{h})$ is the Schlick's approximation of the Fresnel term [Sch94]. $D(\mathbf{h})$ is the distribution function of the microfacets, controlled by the parameters e_x and e_y; the axes of an ellipse that orientates the halfway vector $\mathbf{h}$ of the microfacets, respectively, along the X and Y and thus defining the anisotropy:

$$D(\mathbf{h}) = \frac{\sqrt{(e_x + 1)\,(e_y + 1)}\,(\mathbf{h} \cdot \mathbf{n})^{e_x \cos^2(\phi_h) + e_y \sin^2(\phi_h)}}{4\pi}. \tag{3.7}$$

In order to preserve energy conservation and to model the behavior of the surface's diffuse color near the grazing angle, which disappears due to the increase in specular reflectance, instead of a Lambertian diffuse term, an angle-dependent form of the diffuse component is reported. The expression is based on the consideration that the amount of energy for diffuse scattering is dependent on the Fresnel reflectance at normal incidence:

$$f_{r,d}\,(\mathbf{v_i}, \mathbf{v_r}) = (1 - F(0))\,g(\mathbf{v_i}, \mathbf{v_r}, \mathbf{n})(28\rho_d)/(23\pi), \tag{3.8}$$

where $F(0)$ is the Fresnel reflectance at normal incidence (see Equation (2.10)) and $g(\mathbf{v_i}, \mathbf{v_r}, \mathbf{n}) = [1 - (1 - (\mathbf{n} \cdot \mathbf{v_i})\,/2)^5][1 - (1 - (\mathbf{n} \cdot \mathbf{v_r})\,/2)^5]$. The model is able to describe anisotropic reflections of two layered materials, such as varnished wood for example, and it is physically plausible. A sampling method for Monte Carlo rendering is also provided, based on $D(\mathbf{h})$: it gives the probability density function $p(\mathbf{v_r}) = D(\mathbf{h})/4(\mathbf{v_i} \cdot \mathbf{h})$.

Edwards et al. proposed a framework for transforming the halfway vector $\mathbf{h}$ into different domains to enforce energy conservation but compromising reciprocity [EBJ*06]. By writing Equation (2.11) in terms of $\forall \mathbf{v_r}$ and assuming that it satisfies an equality instead of an inequality, the function

$$Q(\mathbf{v_i}) = f_r(\mathbf{v_i}, \mathbf{v_r})(\mathbf{v_i} \cdot \mathbf{n}) \tag{3.9}$$

can be seen as a probability density function (PDF) over the set of incident directions $\mathbf{v_i}$ on the hemisphere Ω_+. Since the PDF $Q(\mathbf{v_i})$ is related to a PDF $q(\mathbf{h})$ over halfway vectors by the formula $Q(\mathbf{v_i}) = q(\mathbf{h})/(4\mathbf{v_i} \cdot \mathbf{h})$, from Equation (3.9) the following expression for $f_r(\mathbf{v_i}, \mathbf{v_r})$, which conserves energy, can be derived:

$$f_r(\mathbf{v_i}, \mathbf{v_r}) = [q(\mathbf{h})]\,/\,[4(\mathbf{v_i} \cdot \mathbf{n})(\mathbf{v_i} \cdot \mathbf{h})]\,. \tag{3.10}$$

In the space of incident directions $\mathbf{v_i}$ it is difficult to formulate a PDF to describe off-specular reflection and other phenomena. As for the halfway vector domain, near grazing angles the set of allowable halfway vectors changes in a complicated way. If a new domain D_h is defined, together with a PDF $p(\mathbf{l})$ and a bijection $f(\mathbf{h}) = \mathbf{l}$ between the set of halfway vectors $\mathbf{h} \in \Omega_+$ and the set of points $\mathbf{l} \in D_h$, by equating the differential probabilities between D_h and Ω_+, the following can be derived from Equation (3.10):

$$f_r(\mathbf{v_i}, \mathbf{v_r}) = [p(\mathbf{l})d\mu(\mathbf{l})]\,/\,[(4\mathbf{v_i} \cdot \mathbf{h})d\omega_h]\,, \tag{3.11}$$

where $d\mu$ is the differential measure over D_h and $p(l)d\mu(l) = q(\mathbf{h})d\omega_h$. With this framework, a new domain can be defined given $\mathbf{v_r}$, by translating Ω_+ so that the center of its base lies at the tip of $\mathbf{v_r}$. In this way, every point on the translated hemisphere corresponds to an unnormalized halfway vector $\mathbf{h_u} = \mathbf{v_i} + \mathbf{v_r}$. The final step is the transformation of the vectors $\mathbf{h_u}$ to points l on the base of the hemisphere; if the local orientation of the surface is given by the orthogonal vectors $\mathbf{u}$ and $\mathbf{v}$, a point on the disk can be defined by the (u, v) coordinates, hence the PDF $p(l)$ is two-dimensional. A possibility is to scale the halfway vector until its tip lies in the base of the hemisphere, and the resulting energy-conserving BRDF is:

$$f_r(\mathbf{v_i}, \mathbf{v_r}) = \frac{p(l)(\mathbf{v_r} \cdot \mathbf{n})^2}{4(\mathbf{v_i} \cdot \mathbf{n})(\mathbf{v_i} \cdot \mathbf{h})(\mathbf{h} \cdot \mathbf{n})^3} \tag{3.12}$$

since $l = \frac{(\mathbf{v_r} \cdot \mathbf{n})}{(\mathbf{h_u} \cdot \mathbf{n})}\mathbf{h_u}$. This transform allows describing retro-reflective materials by defining a PDF with high values near the center of the disk, which corresponds to a halfway vector in the retro-reflective direction; to specify shiny BRDFs it is enough to define a PDF with high values near the origin of the (u, v) space, which corresponds to $\mathbf{n}$ and gives pure specular reflection. To importance sample the BRDF to obtain $\mathbf{h}$ it is enough to generate a point l on the disk according to $p(l)$ and normalize. Alternatively, the orthogonal projection maps $\mathbf{h_u}$ to the disk along the direction of $\mathbf{n}$: $l = \mathbf{h_u} - (\mathbf{v_i} \cdot \mathbf{n})\mathbf{n}$; it leads to a BRDF with narrower lobes, centered on the direction of perfect reflection. The resulting BRDF, suitable for data fitting, is given by:

$$f_r(\mathbf{v_i}, \mathbf{v_r}) = [1/(4(\mathbf{v_i} \cdot \mathbf{h})^2)]p(l) \parallel \mathbf{v_i} + \mathbf{v_r} \parallel^2 . \tag{3.13}$$

To importance sample the BRDF, once a sample l is generated according to $p(l)$, the unnormalized halfway vector is obtained from the expression $\mathbf{h_u} = l + (\mathbf{v_i} \cdot \mathbf{n})\mathbf{n}$. Within the same framework, two additional BRDF models are described: an empirical, energy-preserving BRDF with limited number of parameters and a BRDF model useful for data fitting, which does not preserve energy.

The Ashikhmin-Premoze model [AP07], or d-BRDF, follows [TS67] and [CT82] microfacet theory and is based on the earlier [AS00] model, with a simplified process of fitting BRDF models to measured data; an efficient sampling technique is also suggested. The Ashikhmin-Premoze model combines an analytic model with a data-driven distribution and also discusses how to fit backscattering measurements to the model [GHP*08]. The model allows the use of an arbitrary normalized function $p(\mathbf{h})$, hence specular highlights can be easily adjusted since their shape depends directly on the distribution. The max term in Equation (3.6), which causes color-banding artifacts as observed in [AP07], is replaced with a smoother term $(\mathbf{v_i} \cdot \mathbf{n}) + (\mathbf{v_r} \cdot \mathbf{n}) - (\mathbf{v_i} \cdot \mathbf{n})(\mathbf{v_r} \cdot \mathbf{n})$. An additional modification is to exclude the $(\mathbf{h} \cdot \mathbf{v_r})$ term to improve the appearance matching with real-world materials. The resulting expression for the specular term, which is reciprocal and non-negative for any non-negative $p(\mathbf{h})$, can be written as:

$$f_{r,s}(\mathbf{v_i}, \mathbf{v_r}) = \frac{F(\mathbf{v_i} \cdot \mathbf{h})\, p(\mathbf{h})k_s}{(\mathbf{v_i} \cdot \mathbf{n}) + (\mathbf{v_r} \cdot \mathbf{n}) - (\mathbf{v_i} \cdot \mathbf{n})(\mathbf{v_r} \cdot \mathbf{n})} \tag{3.14}$$

where k_s is a scaling constant that needs to be chosen in order to fulfill energy conservation. The d-BRDF model improves representation of material reflectance at grazing angles and enables more realistic material appearance. However some effects like retro-reflection cannot be modeled properly. The values of the Fresnel parameters can lie outside the actual range for a given materia, hence they do not have a physical meaning.

3.2   PHYSICALLY BASED MODELS

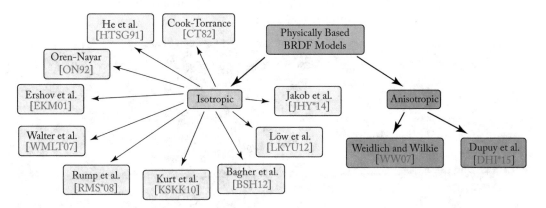

Figure 3.8: Physically based isotropic and anisotropic models described in Section 3.2.

3.2.1   PHYSICALLY BASED MODELS FOR ISOTROPIC MATERIALS

The Cook-Torrance model [CT82] takes into account both specular and diffuse reflections, the latter modeled as Lambertian reflections. As for the specular component, the model assumes that only the fraction of the facets oriented in the direction of h contributes to the final reflection; moreover it accounts for how many facets are visible from different view angles and how they reflect light [WLL*09]. These factors are modeled respectively through the functions D, G, and F:

$$f_{r,s}(\mathbf{v_i}, \mathbf{v_r}) = \frac{F(\theta_r)\, D(\mathbf{h}) G(\mathbf{v_i}, \mathbf{v_r})}{\pi \cos(\theta_r) \cos(\theta_i)}. \tag{3.15}$$

The expression of the distribution $D(\mathbf{h})$ is generally a Gaussian; however other choices are common (see Figure 3.9 for an example with the Beckmann distribution [BS64]): $D(\mathbf{h}) = \cos(\theta_r)\exp^{-\left(\frac{\alpha}{m}\right)^2}$, where α is the angle between $\mathbf{v_i}$ and the reflected $\mathbf{v_r}$ and m is a roughness parameter. The attenuation term G includes both the shadowing and masking effects:

$$G(\mathbf{v_i}, \mathbf{v_r}) = \min\left(1, \frac{2(\mathbf{n} \cdot \mathbf{h})(\mathbf{n} \cdot \mathbf{v_r})}{\mathbf{v_r} \cdot \mathbf{h}}, \frac{2(\mathbf{n} \cdot \mathbf{h})(\mathbf{n} \cdot \mathbf{v_i})}{\mathbf{v_r} \cdot \mathbf{h}}\right). \tag{3.16}$$

One of the important contributions of this work is the formulation of the Fresnel term F, which represents the reflection of polished microfacets, approximated with the following expression:

$$F(\theta) = \frac{(g-c)^2}{2(g+c)^2}\left(1 + \frac{(c(g+c)-1)^2}{(c(g-c)+1)^2}\right), \qquad (3.17)$$

where $c = \mathbf{v_r} \cdot \mathbf{h}$ and $g = \eta^2 + c^2 - 1$, being η the index of refraction. The Cook-Torrance model can properly model metals, plastic with varying roughness, and view-dependent changes in color, although it does not follow the energy conservation principle in the entire hemisphere; additional drawbacks are the not-intuitive parameters.

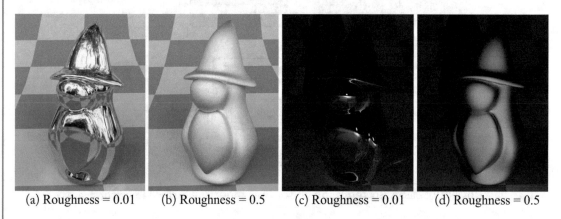

| (a) Roughness = 0.01 | (b) Roughness = 0.5 | (c) Roughness = 0.01 | (d) Roughness = 0.5 |

Figure 3.9: A set of rendering of copper with the Cook-Torrance model [CT82], using the Beckmann distribution. The first two images from the left are rendered under environment map lighting; the last two images on the right under a point light illumination. The surface roughness values are reported in brackets.

The He et al. model [HTSG91] accounts for polarization and masking/shadowing effects, includes specular reflection when the surface roughness is low, and takes into account the nature of light as an electromagnetic wave to model diffraction and interference, thus resulting in a very complex BRDF model. The model is able to represent metal, non-metal and plastic with smooth and rough surfaces, and all parameters are physically based. The contribution to the reflection is given by three components, namely the specular term for mirror-like reflections, the directional diffuse, and the uniform diffuse. The specular term describes mirror-like reflections from the mean plane of the surface:

$$f_{r,s}(\mathbf{v_i}, \mathbf{v_r}) = \frac{|F(\theta_r)|^2 \exp(-g(\sigma, \lambda))S(\mathbf{v_i}, \mathbf{v_r})}{\cos(\theta_i)d\omega_i}\Delta, \qquad (3.18)$$

where F is the Fresnel reflectivity, S is a shadowing function, σ refers to the surface roughness, λ is the wavelength of the incident light, Δ is a delta dirac function equal to 1 in the specular cone of reflection, and $d\omega_i$ is the incident solid angle. The function of the surface roughness $g(\sigma, \lambda)$ is given by the expression $g(\sigma, \lambda) = (((2\pi\lambda)/\sigma)(\cos(\theta_i) + \cos(\theta_r)))^2$. For a smooth surface $S \to 1$ and $g \to 0$, hence the expression of the specular term becomes the specular reflectivity of a specular surface. As for the diffuse directional term, it describes diffraction and interference effects, which spread out the reflected field over the hemisphere, with a possible directional and non-uniform shape of the light intensity distribution:

$$f_{r,dd}(\mathbf{v_i}, \mathbf{v_r}) = \frac{F(\mathbf{b}, \mathbf{p}) S \tau^2}{\cos(\theta_r)\cos(\theta_i)16\pi} \sum_{m=1}^{+\infty} \frac{g^m e^{-g(\sigma,\lambda)}}{m!m} e^{\left(-\frac{w_y^2 \tau}{4m}\right)}. \tag{3.19}$$

The directional diffuse reflection depends on surface roughness σ and on the autocorrelation length τ. The other parameters are the bisecting unit vector $\mathbf{b}$, the incident polarization state vector $\mathbf{p}$ and the wave vector change $\mathbf{w_v}$. For very smooth surfaces, $f_{r,dd}$ decreases to zero, and for slightly rough surfaces the maximal values are aligned with the specular direction. As the roughness is increased, the maximal values progressively move from off-specular angles to grazing angles for very rough surfaces. The uniform diffuse term is approximated with a Lambertian model and denoted by $f_{r,ud}(\mathbf{v_i}, \mathbf{v_r}) = a(\lambda)$. An experimental analysis reported in [NDM05] indicates that when polarization and spectral dependencies are omitted, the He et al. model does not produce noticeably better visual results than the Cook-Torrance model [CT82]. The model does not suggest a sampling method and does not describe anisotropic materials.

Oren-Nayar [ON94] enhanced the Lambertian model for rough diffuse surfaces to describe in a more realistic way the behavior of real-world materials like concrete, sand, and cloth, which show increasing brightness as the viewing direction approaches the light source direction, rather than being independent of the viewing direction. A rough diffuse surface is modeled as a collection of long symmetric V-cavities, each of which consists of two microfacets with a Lambertian reflectance; microfacets orientated toward the light source diffusely reflect some light back to the light source (backscatter). The model takes into account masking, shadowing, and inter-reflections. The expression is given by:

$$f_r(\mathbf{v_i}, \mathbf{v_r}) = \frac{\rho_d}{\pi}(A + B \max(0, \cos(\phi_i - \phi_r))\sin(\alpha)\tan(\beta)), \tag{3.20}$$

where $\alpha = \max(\theta_r, \theta_i)$; $\beta = \min(\theta_r, \theta_i)$; given the surface roughness σ, the expressions for A and B are: $A = 1 - [(0.5 \cdot \sigma^2)/(\sigma^2 + 0.33)]$; $B = (0.45 \cdot \sigma^2)/(\sigma^2 + 0.09)$. This model, widely used in computer graphics, obeys the reciprocity principle and reduces to the Lambertian model when $\sigma = 0$. In Figure 3.10 we report an example depicting a terracotta pot; notice the differences around the edges (Figure 3.10c).

The multilayered model by Ershov et al. [EKM01] represents car paint and consists of binder pigment particles, flakes, and flake coatings. The model approximates the BRDF of each sub-layer and then merges sub-layers together and is able to produce a realistic appearance for car paints and models their components (binder, pigment particles, flakes). However, due to

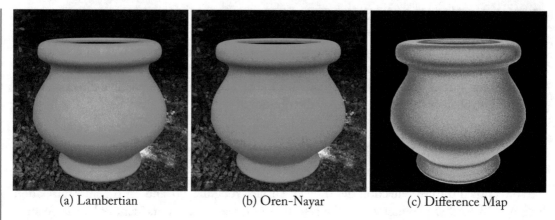

| (a) Lambertian | (b) Oren-Nayar | (c) Difference Map |

Figure 3.10: Oren-Nayar model. In (a) a terracotta rendered with a Lambertian model; in (b) the same scene is rendered using the Oren-Nayar model, assuming a very rough surface; in (c) a false-color difference map between (a) and (b) is reported.

the complexity of the layered model, the computational time is significantly high. An updated version of the model is simplified to a bi-layered model and presents a substrate layer as a solid paint film where the reflectance is Lambertian and a transparent binder layer with embedded flakes (see Figure 3.11). Flakes are considered as partially transparent colored mirrors, with the assumption that the reflectance of flakes does not depend on the incident direction and inter-reflections between flakes, so their interaction with light is modeled using constant reflectance values [EDKM04]. A large number of parameters are required and not all of them can be directly measured. This model is suitable for interactive design of automotive paints by solving through optimization the problem of finding pigment composition of a paint from its BRDF.

Figure 3.11: Composition of a two-layer paint used in the Ershov et al. models [EDKM04].

Walter et al. [WMLT07] extend the microfacets theory introduced by [CT82] to simulate transmission through etched glass and other rough surfaces, thus taking into account the BSDF. The work by Smith [Smi67], which investigated the geometrical self-shadowing of a surface described by Gaussian statistics, is also extended by deriving a shadowing function from any microfacet distribution D; the BRDF component follows (3.1). The distribution D is different from previous models and has been developed to better fit measured data; it is named GGX and has the following expression:

$$D(\mathbf{h}) = \frac{\alpha_g{}^2 \chi^+(\mathbf{h} \cdot \mathbf{n})}{\pi \cos^4 \theta_h \left(\alpha_g{}^2 + \tan^2 \theta_h\right)^2}, \qquad (3.21)$$

where $\alpha_g{}^2$ is a width parameter and $\chi^+(x)$ is equal to one if $x > 0$ and zero if $x \leq 0$. The GGX distribution has a stronger tail than commonly used distributions, such as Beckmann and Phong, and thus tends to have more shadowing; in [BSH12] it has been observed that the GGX distribution is identical to the Trowbridge-Reitz distribution [TR75]. From D it is possible to derive a simple sampling equation and the expression of G, which is given by:

$$G(\mathbf{v}_i, \mathbf{v}_r, \mathbf{h}) \approx G_1(\mathbf{v}_i, \mathbf{h}) G_1(\mathbf{v}_r, \mathbf{h}) \qquad (3.22)$$

$$G_1(\mathbf{v}_x, \mathbf{h}) = \chi^+\left(\frac{\mathbf{v}_x, \mathbf{h}}{\mathbf{v}_x, \mathbf{n}}\right) \frac{2}{1 + \sqrt{1 + \alpha_g{}^2 \tan^2 \theta_x}}. \qquad (3.23)$$

As previously observed, the GGX distribution has a longer tail than other distributions (see Figures 3.12 and 3.13). However the GGX distribution fails to properly capture the glowy highlights of highly polished surfaces like the chrome sample in the MERL database [MPBM03a], with a narrow specular peak and a much wider specular tail [MHH*12]. An anisotropic extension of the distribution, named Generalized-Trowbridge-Reitz, has been proposed by Burley [MHH*12]; a symmetric extension of the GGX to the entire ellipsoid domain, suitable for volumetric anisotropic materials, is described by Heitz et al. [HDCD15].

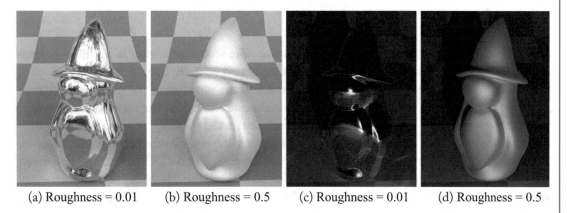

(a) Roughness = 0.01   (b) Roughness = 0.5   (c) Roughness = 0.01   (d) Roughness = 0.5

Figure 3.12: A set of rendering of silver with the GGX model [WMLT07]. The first two images from the left are rendered under environment map lighting, the two images on the right under a point light illumination. The surface roughness values are reported in brackets.

Another method aimed to describe the complex reflectance behavior of a car paint is described in Rump et al. [RMS*08], which represents the reflectance with the first hybrid an-

(a) Cook-Torrance   (b) GGX

Figure 3.13: Visual comparison about the effect of the microfacets distribution between the Cook-Torrance (on the left) and GGX (on the right) models. For similar values of surface roughness, the GGX model shows a wider specular tail. The metal rendered in the images is iridium.

alytical BRDF and image-based BTF representation; the acquisition setup is described in Section 4.4. The appearance of metallic car paint is separated into the homogeneous BRDF part, which describes the reflection behavior of the base and the top layer of the paint, and the spatially varying BTF part, which is caused by the aluminium flakes. The homogeneous part is represented by a multi-lobe version of the Cook Torrance model [CT82]. In order to account for the characteristics of pearlescent paint, which show view-dependent off-specular color changes, the model includes a spectral view and light-dependent part. The BRDF parameters are derived from the BTF measurements by means of a fitting procedure; the BRDF is calculated for every pixel and subtracted in the RGB space from the captured images. The resulting images contain only flakes data and are used for a copy-and-paste synthesis approach.

Kurt et al. [KSKK10] proposed a BRDF model based on the halfway vector representation and Beckmann distribution. The model is physically plausible, can represent anisotropic materials, can accurately fit data, and suggests an efficient importance sampling method, based on the strategy proposed by Ward et al. but with a different weighting function, which makes it particularly suitable for Monte Carlo Rendering algorithms. The basic BRDF model they propose is the sum of a pure Lambertian term and a single specular lobe, which can be readily extended to multiple specular lobes representation, to model mixture materials like a car paint:

$$f_r\left(\mathbf{v_i}, \mathbf{v_r}\right) = \frac{k_d}{\pi} + \sum_{l=1}^{N} \frac{k_{sl} F_l(\mathbf{v_r}, \mathbf{h}) D_l(\mathbf{h})}{4(\mathbf{v_r} \cdot \mathbf{h})\left((\mathbf{v_i} \cdot \mathbf{n})(\mathbf{v_r} \cdot \mathbf{n})\right)^{\alpha_l}},$$ (3.24)

where N is the number of lobes, k_d is the diffuse albedo, k_{sl} is the specular reflectivity per lobe, F_l is a per-lobe Fresnel term, D_l a per-lobe normalized microfacet distribution, and α_l is a set of parameters that needs to be chosen carefully to enforce energy conservation.

Bagher et al. suggested a function of $\tan^2 \theta_h^{-p}$ for the distribution D, where p depends on the model [BSH12], in order to enhance data fitting for single-layered materials like metals, metallic paints, and shiny plastics. Otherwise it is very difficult to fit with commonly used distributions and generally requires several lobes due to the shape of the decrease in their BRDFs, close to exponential at large angles but sharper at small angles. The model presented is the Cook–Torrance [CT82], in which the microfacets distribution is designed to efficiently and accurately approximate measured data. The distribution resulting from the suggested slope is called SGD (Shifted Gamma Distribution):

$$D(\theta_h) = \frac{\chi_{[0,\pi/2]}(\theta_h)\alpha^{p-1}e^{-\frac{\alpha^2+\tan^2 \theta_h}{\alpha}}}{\pi \cos^4 \theta_h \Gamma(1-p,\alpha)\left(\alpha^2 + \tan^2 \theta_h\right)^p},$$ (3.25)

where α is a fitting parameter, $\chi_{[0,\pi/2]}(\theta_h)$ is is equal to 1 if $\theta_h < \pi/2$ and 0 otherwise, and Γ is the incomplete Gamma function:

$$\Gamma(1-p,\alpha) = \int_{\alpha}^{\infty} t^{-p}e^{-t}dt.$$ (3.26)

From the SGD it is possible to derive the shadowing function G and a sampling method.

Low et al. [LKYU12] proposed two isotropic models for glossy surfaces, based either on the Rayleigh-Rice light scattering theory (smooth surface BRDF) or the microfacet theory (microfacet BRDF). Both models make use of a modified version of the ABC model [CTL90], [CT91], which was originally formulated to fit the Power Spectral Density (PSD) of some measured smooth surfaces. The PSD describes the surface statistics in terms of the spacial frequencies f_x and f_y, which depend on the wavelength λ of the incident light:

$$f_x\left(\mathbf{v_i}, \mathbf{v_r}\right) = \left(\sin \theta_r \cos \phi_r - \sin \theta_i\right)/\lambda;$$ (3.27)

$$f_y\left(\mathbf{v_i}, \mathbf{v_r}\right) = \left(\sin \theta_r \sin \phi_r\right)/\lambda.$$ (3.28)

The ABC model [CTL90], [CT91] is able to model the inverse power law shape PSD of polished data, and it is given by:

$$PSD(f) = A'/\left(1 + B^2 f^2\right)^{\frac{C+1}{2}},$$ (3.29)

where A is determined by low-frequency spectral density, $B = 2\pi l_0$, l_0 is the autocorrelation length, $C > 0$, $f = \sqrt{f_x^2 + f_y^2}$, $A' = \Gamma((c + 1)/2)AB/[2\Gamma(c/2)\sqrt{\pi}]$, and Γ is the gamma function. In [LKYU12] the ABC model is simplified to $S(f) = a/(1 + bf^2)^c$, where the mapping of the new parameters to the original ABC is: $a = A'$, $b = B^2$, and $c = (C + 1)/2$; in practice, narrower specular peaks are obtained by increasing b, whereas c controls the fall-off rate of wide-angle scattering. The smooth surface BRDF has the following expression:

$$f_r(\mathbf{v_i}, \mathbf{v_r}) = (k_d/\pi) + O\,F(\theta_d)\,S\left(\|\mathbf{D_p}\|\right), \tag{3.30}$$

where k_d is a scaling factor for the Lambertian term, O is a modified obliquity factor, and $F(\theta_d)$ is the Fresnel term in Equation (3.15) with extinction coefficient set to zero and aimed to approximate the reflectivity polarization factor, which depends on the surface material properties. $\mathbf{D_p}$ is the projected deviation vector, defined as $\mathbf{D_p} = \mathbf{v_{r,p}} - \mathbf{r_{v_i,p}}$, where $\mathbf{v_{r,p}}$ is the projection of $\mathbf{v_r}$ on the surface tangent plane and $\mathbf{r_{v_i_p}}$ is the projection of the mirror direction of $\mathbf{v_i}$ on the surface tangent plane. To deal with unreliable data near grazing angles, the value suggested for the obliquity factor O is 1 instead of the typical definition of $O = \cos\theta_i \cos\theta_r$. The microfacet model is based on Cook–Torrance [CT82] and makes use of the modified ABC distribution:

$$f_r(\mathbf{v_i}, \mathbf{v_r}) = \frac{k_d}{\pi} + \frac{F(\theta_h)S\left(\sqrt{1 - \mathbf{h}\cdot\mathbf{n}}\right)G(\mathbf{v_i}, \mathbf{v_r})}{\mathbf{v_i}\cdot\mathbf{n}\,\mathbf{v_r}\cdot\mathbf{n}}, \tag{3.31}$$

where F and G are the same as in Equation (3.15), S is the modified ABC distribution, and k_d is again a scaling factor for the diffuse component; the parameter a of S is used as a scaling factor for the specular term, hence the distribution is not normalized. The model is reciprocal but does not obey energy conservation. Both models provide accurate fits to measured data (see the blue metallic paint in Figure 3.14), with the microfacet model showing lower errors, and accurately represent scattering from glossy surfaces with sharp specular peaks and non-Lambertian wide angle scattering. For both models an efficient importance sampling strategy is suggested.

The discrete stochastic model by Jakob et al. [JHY*14] extends the microfacet theory by replacing the continuous distribution of microfacets in the Cook-Torrance model [CT82] with a discrete one, thus assuming that a surface consists of a high but finite number of scattering particles. This assumption facilitates modeling a controllable, non-smooth spatially varying BRDF appearance of a glittery surface, like mica flakes, ice crystals, metallic car paint, and craft glitter for decorations. The notion of multiscale BRDF is introduced, which takes into account finite areas and solid angles rather than single points and directions:

$$f_r(A, \mathbf{v_i}, \omega_r) = \frac{(\mathbf{v_i}\cdot\mathbf{h})\,F(\mathbf{v_i}\cdot\mathbf{h})\,D(A, \omega_h),\,G(\mathbf{v_i}, \mathbf{v_r}, \mathbf{n})}{a(A)\sigma(\omega_r)(\mathbf{v_i}\cdot\mathbf{n})(\mathbf{v_r}\cdot\mathbf{n})}, \tag{3.32}$$

where A is the area around the point p into account, $a(A)$ its surface area, $\omega_h :=$ $\{(\mathbf{v_i} + \mathbf{v_r}/\|\mathbf{v_i} + \mathbf{v_r}\|), \mathbf{v_r} \in \omega_r\}$ is the set of microfacet normals that reflect from $\mathbf{v_i}$ into the finite

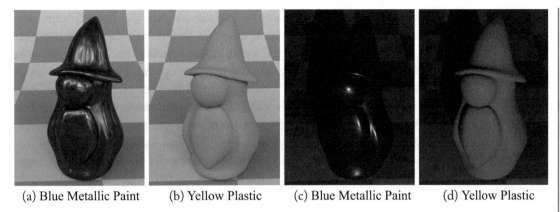

(a) Blue Metallic Paint   (b) Yellow Plastic   (c) Blue Metallic Paint   (d) Yellow Plastic

Figure 3.14: Low et al. [LKYU12] microfacets BRDF model. The first two images from the left depict a blue metallic paint and a yellow plastic, rendered under environment map lighting. The last two images on the right refer to the same materials, rendered under point light illumination.

solid angle ω_r around $\mathbf{v_r}$, $\sigma(\omega_r)$ is the area of ω_r on the unit sphere, F is the Fresnel term, and G models shadowing and masking. The discrete multiscale microfacets distribution D is defined as:

$$D(A, \omega_h) = \frac{1}{N} \sum_{k=1}^{N} 1_{\omega_h}(\mathbf{v_h}^k) 1_A(p^k), \qquad (3.33)$$

where p^k and $\mathbf{v_h}^k$ are the position and normal of the kth microfacet of a list of N microfacets, 1_A and 1_{ω_h} are the indicator functions of the sets A and ω_h, respectively. The indicator functions control the appearance of the surface, since they determine which microfacets in A reflect light into the solid angle ω_r around $\mathbf{v_r}$: a high number of participating facets gives a smoother appearance than a low number, which gives instead a strongly glittery appearance. An efficient implementation of the model is discussed, together with an importance sampling strategy for Monte Carlo renderers.

3.2.2   PHYSICALLY BASED MODELS FOR ANISOTROPIC MATERIALS

To simulate both smooth and rough multi-layered materials, Weidlich and Wilkie [WW07] proposed to combine several microfacet based layers into a single physically plausible BRDF model. Their model assumes that any microfacet is large in relation to the layer thickness, models the absorption of part of the light when it travels inside a transparent material, and includes a total reflection term, when light propagates at an angle of incidence greater than the critical angle; the simplicity of the model does not allow reproducing effects like iridescence.

Dupuy et al. [DHI*15] proposed an approach to automatically convert an arbitrary material to a microfacet BRDF. The facet distribution is obtained by solving an eigenvector problem, based solely on backscattering samples and simplifying the Fresnel term to a constant; once an eigenvector with all positive components is found with the power iteration method, its values are linearly interpolated to build a continuous distribution. The Fresnel term is then recovered by calculating for each color channel the average ratio between the input and an ideal mirror microfacet BRDF with a constant Fresnel term equal to 1, in the form $f_r(v_i, v_r) = D(\mathbf{h})G(v_i, v_r)/(4\cos\theta_i\cos\theta_r)$. This method can fit an anisotropic BRDF in a few seconds (on an Intel 2.4 GHz Core i5 CPU, when the number of samples evaluated from the material in the elevation and azimuthal directions is up to 4,096) and allows editing the properties of the roughness distribution. However its accuracy depends on the density of the measurements in the backscattering direction, thus limiting the applicability to more complex anisotropic BRDFs [FHV15].

3.3   DATA-DRIVEN MODELS

The general idea behind this class of models is to use the measured data to derive some ad hoc model, generally based on the principle that a continuous function can be represented by a linear combination of basis functions, and a mixture of basis functions can be used for interpolation: for instance, in the Fourier basis, functions are expressed as a sum of sinusoidal and cosinusoidal terms, whereas in the polynomial basis, a collection of quadratic polynomials are used with real coefficients.

A possible way to represent BRDFs is to project them onto an orthonormal basis [AMHH08], mapped onto a unit disc and projected on to a hemisphere. Spherical wavelets, spherical harmonics, and Zernike polynomials are mathematically and computationally efficient, since they represent the shape of the BRDF as the sum of low and high frequency functions to capture the shape of the BRDF. Wavelets [LF97, CPB03, CBP04] can represent large specular peaks more efficiently than spherical harmonics [WAT92] and Zernike polynomials [Rus98]. The limitation of these methods is the significant memory requirement even to obtain simple BRDFs, since a large number of basis functions is generally required.

Separable decompositions of a high-dimensional function f can be used to approximate it to arbitrary accuracy, using a sum of products of lower-dimensional functions. Four dimensional BRDFs can be written as a sum of terms, each of which is the product of two-dimensional functions $g(\cdot, \cdot)$ and $h(\cdot, \cdot)$:

$$f_r(v_i, v_r) = f_r(\theta_i, \phi_i, \theta_r, \phi_r) \approx \sum_{k=1}^{N} g_h(\theta_i, \phi_i)h_k(\theta_r, \phi_r). \qquad (3.34)$$

This representation directly approximates the fully tabulated BRDF over all directions, implementing manageability of data for use in rendering systems, and it is also useful for the purpose of importance sampling. If a good approximation can be found for a small N, a separable decomposition is capable of high compression rates, thus resulting in a compact way to store large measured datasets while maintaining accurate representation. The parameterization of the lower-dimensional functions can improve the performance of the decomposition and needs to be wisely chosen in order to minimize the number of functions needed for BRDF representations. A common reparameterization makes use of the angle halfway between the incident and exitant directions and the difference angle [Rus98] (see Figure 2.4):

$$f_r(\mathbf{v_i}, \mathbf{v_r}) \approx \sum_{k=1}^{n} g_k(\mathbf{v_h}) h_k(\mathbf{v_d}) \tag{3.35}$$

where $\mathbf{v_h}$ and $\mathbf{v_d}$ arise from the reparameterization.

A common technique to obtain a separable representation is the Singular Value Decomposition (SVD) [PFTV88]. Given a matrix M, its SVD is the factorization in the form $M = USW^T$, where $S = diag(\sigma_k)$ is a diagonal matrix of singular values; the columns of $U = [\mathbf{u}_k]$ and $W = [\mathbf{w}_k]$ are orthonormal. The matrix M can be written as:

$$M = USW^T = \sum_{k=1}^{K} \sigma_k \mathbf{u}_k \mathbf{w}^T_k \tag{3.36}$$

where $\mathbf{u}_k \mathbf{w}^T_k$ is an outer product. In Fournier [Fou95] the SVD decomposition is used to approximate the Blinn-Phong BRDF using Ward's measured data; the BRDF is approximated with a sum of terms, each of which is the product of two functions, one of the incident and one of the outgoing direction.

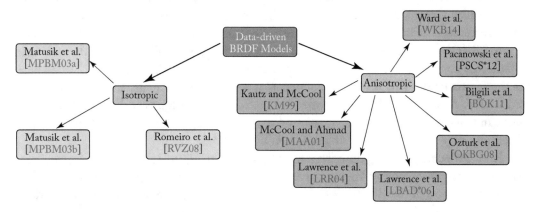

Figure 3.15: Phenomenological isotropic and anisotropic models described in Section 3.3.

3.3.1   DATA-DRIVEN MODELS FOR ISOTROPIC MATERIALS

Matusik et al. [MPBM03a] presented a set of data-driven reflectance models, based on either linear and non-linear dimensionality reduction. A set of 104 isotropic BRDFs, parameterized using the half-angle [Rus98], are discretized into $90 \times 90 \times 180$ bins that are smoothed by removing outliers. The linear dimensionality reduction used is the Principal Component Analysis (PCA), which allows determination of a set of basis vectors that span the linear subspace on which the BRDFs lie. The RGB color channels are assembled together and analyzed in the log space in order to reduce the difference between specular and non-specular values. A linear combination of a subset of the principal components is used for the reconstruction, and in most cases 30–40 components give good results. As for the non-linear dimensionality reduction, the charting algorithm [Bra02] has been used since it gives good results even with a small number of samples and at the same time reduces the noise in the data. The idea behind charting is that data lies on a low-dimensional manifold embedded in the sample space and tries to find a kernel-based mixture of linear projections to smoothly map the samples on the coordinate system while preserving local relationships between the sample points. Since each dimension performs a noise suppression in a different direction, the error does not decrease monotonically. However, they found that the BRDF data lies in a 10D manifold, and a 15D manifold would suffice to synthesize new BRDFs even over long distances. The advantages of such data-driven BRDF models are the realistic appearance and meaningful parameterization.

Matusik et al. performed a wavelet analysis for all of the measured isotropic BRDFs [MPBM03b] in order to find the maximum required frequency to sample any arbitrary BRDF correctly. For each BRDF a non-uniform wavelet transform is applied to determine the highest coefficients able to reconstruct the BRDF itself with high precision while setting to zero the rest of the coefficients. The union of the sets of non-zero wavelet coefficients (about 69,000), which show some degree of coherence among different BRDFs, corresponds to a set of wavelet functions called Common Wavelet Basis (CWB). The CWB allows reconstruction of a BRDF by solving a system of linear equations. In the same work, a simple approach to represent a new measured BRDF is presented, using a linear combination of the BRDFs in the dataset. Using this data they construct an over-constrained system of equation in the form $P \times C \approx B$, where P is the matrix of the BRDFs in the dataset, C a vector of coefficients, and B the new measured BRDF. A subset X of the rows of P is constructed in such a way the ratio between the highest and lowest eigenvalue of the matrix $X^T X$ is small. Experimentally they have found that 800 samples are enough to represent a new BRDF.

Romeiro et al. [RVZ08] describe a method for inferring the reflectance of isotropic materials from images, assuming known curved surface with known natural illumination. To reduce the dimension of the BRDF domain, parameterized using the halfway vector and difference angle [Rus98], thanks to the reciprocity assumption it is possible to apply the projection $\phi_d \leftarrow \phi_d + \pi$; for isotropic materials it is possible to apply the projection onto the domain $(\theta_h, \theta_d, \phi_d)$, and in case of bilateral symmetry (i.e., if the reflectance of the material shows little changes when $\mathbf{v_r}$ is reflected about the incident plane), it is possible to apply the projection $\phi_d \leftarrow \phi_d + \pi/2$. If a material is bivariate, in other words it satisfies a further generalization of isotropy, bilateral symmetry and reciprocity, the projection onto the domain $(\theta_h, \theta_d) \in [0, \pi/2]$

is allowed. A bivariate representation is often sufficient to capture off-specular reflections, retro-reflection, and Fresnel effect. Under these assumptions, the resulting 2D domain is sampled using the functions $s(\theta_h, \theta_d) = 2\theta_d/\pi$ and $t(\theta_h, \theta_d) = \sqrt{2\theta_h/pi}$, which allow to increase the sampling density near specular reflections; since bivariate BRDFs vary slowly over a significant region of their domain, an optimization framework with a smoothness constraint is employed to recover the BRDF:

$$\underset{f_r \geq 0}{\text{argmin}} \parallel I - Lf_r \parallel_2^2 + \alpha \left(\parallel \Lambda_s^{-1} D_s f_r \parallel_2^2 + \parallel \Lambda_t^{-1} D_t f_r \parallel_2^2 \right) \tag{3.37}$$

where $\mathcal{S} = (s_i, t_i)$ is a uniform grid in the BRDF domain, D_s and D_t are $|\mathcal{S}| \times |\mathcal{S}|$ derivative matrices, α is a regularization parameter, Λ_s and Λ_t are $|\mathcal{S}| \times |\mathcal{S}|$ matrices that control non-uniform regularization in the (θ_h, θ_d) domain, and L is a lighting matrix. The term I is related to the rendering equation:

$$I(\mathbf{v_r}, \mathbf{n}) = \int_\Omega L(R_n^{-1} \mathbf{v_i}) f_r(s(\mathbf{v_i}, R_n \mathbf{v_r}), t(\mathbf{v_i}, R_n \mathbf{v_r})) \cos \theta_i \, dv_i, \tag{3.38}$$

where R_n rotates $\mathbf{n}$ toward the z-axis and $\mathbf{v_r}$ toward the xz-plane. In order to obtain good results, the environment illumination used to capture the 2D picture must allow sufficient observations of the BRDF (θ_h, θ_d) domain, in particular in regions corresponding to specular reflections, retro-reflections, and grazing angles, which occur respectively at $(\theta_h \approx 0)$, $(\theta_d \approx 0)$ and $(\theta_d \approx \pi/2)$.

3.3.2 DATA-DRIVEN MODELS FOR ANISOTROPIC MATERIALS

A technique for separable decomposition of BRDFs based on either SVD (Equation (3.41)) or Normalized Decomposition (ND) (Equation (3.42)) is described by Kautz and Mc-Cool [KM99]. In both cases the separable decomposition $f_{sd,r}$ to approximate the BRDF f_r has the form:

$$f_r(\mathbf{v_i}, \mathbf{v_r}) = f_{sd,r}(\mathbf{P_x}(\mathbf{v_i}, \mathbf{v_r}), \mathbf{P_y}(\mathbf{v_i}, \mathbf{v_r})), \tag{3.39}$$

where P_x and P_y are vector functions. In the following, the parameters of $f_{sd,r}$ are $\mathbf{x} = \mathbf{P_x}(\mathbf{v_i}, \mathbf{v_r})$ and $\mathbf{y} = \mathbf{P_y}(\mathbf{v_i}, \mathbf{v_r})$. The matrix M of Equation (3.36) consists of the tabulated and reparameterized BRDF values of $f_{sd,r}(\mathbf{x}, \mathbf{y})$:

$$M = \begin{pmatrix} f_{sd,r}(\mathbf{x}_1, \mathbf{y}_1) & \cdots & f_{sd,r}(\mathbf{x}_1, \mathbf{y}_K) \\ \vdots & \ddots & \vdots \\ f_{sd,r}(\mathbf{x}_K, \mathbf{y}_1) & \cdots & f_{sd,r}(\mathbf{x}_K, \mathbf{y}_K) \end{pmatrix}. \tag{3.40}$$

The resulting $\mathbf{u}_k$ and $\mathbf{w}_k$ from the SVD of M can be interpolated in order to obtain the 2D functions $u_k(\mathbf{x})$ and $w_k(\mathbf{y})$:

$$f_{SVD,r}(\mathbf{x}, \mathbf{y}) \approx \sum_{k=1}^N \sigma_k u_k(\mathbf{x}) w_k(\mathbf{y}). \tag{3.41}$$

As for the ND factorization:

$$f_{ND,r}(\mathbf{x}, \mathbf{y}) \approx g_1(\mathbf{x}, \mathbf{y}) h_1(\mathbf{x}, \mathbf{y}), \tag{3.42}$$

where g_1 is a constant if $\mathbf{P_x}(\mathbf{v_i}, \mathbf{v_r})$ is fixed and scales the profile h_1, while SVD can produce optimal approximations and minimizes the RMS error, it is expensive in terms of time and space resources and can produce negative factors in the expansion; the ND algorithm does not guarantee optimality but requires less memory than the SVD decomposition and is faster. The lower-dimensional functions are stored into texture maps to allow multiplications being performed by compositing or multitexturing.

McCool and Ahmad presented [MAA01] a decomposition algorithm for both isotropic and anisotropic BRDFs. The algorithm is based on logarithmic homomorphism (Equation (3.43)) and is general enough to approximate BRDFs with an arbitrary number of positive factors and degree of precision while satisfying the Helmholtz reciprocity but is limited to point and directional light sources. The authors describe a simple parameterization (Equation (3.44)) and demonstrate that it is possible to limit the storage cost to just two texture maps, obtaining good compression ratios:

$$\log\left(f_r(\mathbf{v_i}, \mathbf{v_r})\right) \approx \sum_{j=1}^{N} \log\left(p_j\left(\pi_j(\mathbf{v_i}, \mathbf{v_r})\right)\right), \tag{3.43}$$

$$f_r(\mathbf{v_i}, \mathbf{v_r}) \approx p(\mathbf{v_i})q(\mathbf{h})p(\mathbf{v_r}) \tag{3.44}$$

where $p(\cdot)$ are two-dimensional functions and π_j are projection functions $\mathbb{R}^4 \rightarrow \mathbb{R}^2$. The logarithmic transformation tends to disregard large peaks in the data and smoothes specular highlights, which may lead to high approximation errors.

Lawrence et al. [LRR04] presented an importance sampling algorithm for arbitrary BRDFs based on reparameterizing the BRDFs using the half-angle or the incident angle, followed by a non-negative matrix factorization, which is essential for sampling purposes:

$$f_r(\mathbf{v_i}, \mathbf{v_r})(\mathbf{v_i} \cdot \mathbf{n}) \approx \sum_{j=1}^{J} F_j(v_r) \sum_{k=1}^{K} u_{jk}(\theta_p)v_{jk}(\phi_p). \tag{3.45}$$

The factored form (Equation (3.45)) allows for expressing the BRDF, multiplied by the cosine of the incident angle, as a sum of a small number of terms, each of which is a product of a 2D function F_j dependent only on the outgoing direction and two 1D functions u_{jk}, v_{jk} dependent on the angle chosen for the parameterization $\mathbf{v_p} = (\theta_p, \phi_p)$. The 1D functions are used to interpret the factors as 1D probability distributions. The results are generally accurate and the technique can be used for sampling BTFs and light fields but does not enforce reciprocity, and the representation may present a discontinuity at the pole of the angle selected for the parameterization.

In later work Lawrence et al. [LBAD*06] presented an algorithm based on linear constraint least squares, capable of compact and accurate SVBRDF representation for rendering. Under the assumption that BRDFs are blended linearly over the surface, the matrix factorization algorithm provides an editable decomposition and can represent directional and spatial reflectance behavior of a material. The described Inverse Shade Tree (IST) representation takes as input a measured materials dataset and a user-supplied tree structure and fills in the leaves of the tree. IST proceeds top-down at each stage, decomposing the current dataset according to the type of node encountered in the tree. The leaves provide editability since they correspond to pieces that are meaningful to the user. The Alternating Constrained Least Squares algorithm (ACLS) decomposes the SVBRDF into basis BRDFs as 4D functions in tabular form, which are then decomposed into 2D functions and further into 1D curves. The 1D curves represent data simply and accurately for isotropic materials; for anisotropic materials the decomposition ends into 2D functions. The advantage of ACLS is the possibility to easily add linear constraints, thus allowing to enforce energy conservation, reciprocity, and monotonicity, other than sparsity and non-negativity, but it requires building a regularly sampled data matrix for factorization.

Tensor representation has been previously used for interactive modification of the material properties and relighting by Sun et al. [SZC*07]. Based on the observation that high-frequency specular lobes generally require a large number of basis terms for reconstruction, thus precluding interactive performance, the BRDFs are separated into a specular lobe $f_{r,s}$ and the remainder $f_{r,rm}$. The specular lobe $f_{r,s}$ is modeled as a sum of four Gaussians, with different neighborhood support (0° for perfect mirror reflection, 7°, 14°, and 21° for broader Gaussians). By removing $f_{r,s}$, the BRDF is left with mainly low frequency terms that can be modeled with a small basis by tensor approximation.

When a non-linear function is used for BRDF data fitting, there are several shortcomings, due to the number of parameters that can be large depending on the model and the number of lobes, and to the non-linear estimation process, which can be computationally expensive. Ozturk et al. [OKBG08] proposed a representation based on response surface models, defined as a polynomial function of order p in k variables, and expressing BRDFs as functions of the incoming and outgoing direction and transforming the variables of some non-linear reflectance models (specifically Ward [War92], Lafortune [LFTG97], and Ashikhmin-Shirley [AS00], described in Section 3.1.2) using Principal Component Analysis, thus obtaining a linear representation. This reciprocal but not energy-preserving representation is general enough to model both isotropic and anisotropic materials, diffuse and glossy.

Bilgili et al. [BÖK11] proposed to represent four-dimensional measured BRDFs data as a function of tensor products, factorized using Tucker decomposition [Tuc66], a generalization of higher order Principal Component Analysis. Tensors are a generalization of scalars and vectors to higher orders, and their rank is defined by the number of directions; for example, a scalar is a zero-order tensor and a vector a first-order tensor; the Tucker factorization decomposes a tensor into a set of matrices and one small core tensor. The logarithmic transformation of a 4D BRDF data matrix $B = b_{ijkl}$, based on the halfway vector representation can be roughly approximated by setting all the Tucker parameters to 1:

$$\log(b_{ijkl}) \approx g f_1(\theta_{hi}) f_2(\phi_{hj}) f_3(\theta_{rk}) f_4(\phi_{rl}), \tag{3.46}$$

where $i = 1, \ldots, N_{\theta_h}$, $j = 1, \ldots, N_{\phi_h}$, $k = 1, \ldots, N_{\theta_r}$, $l = 1, \ldots, N_{\phi_r}$, and N_{θ_h}, N_{ϕ_h}, N_{θ_r}, N_{ϕ_r} are the sampling resolution of the BRDF data, g is the zero-order core tensor, $f_1(\theta_{hi})$, $f_2(\phi_{hj})$, $f_3(\theta_{rk})$, $f_4(\phi_{rl})$ are univariate tensor functions respectively evaluated at θ_{hi}, ϕ_{hj}, θ_{rk}, and ϕ_{rl}; the logarithmic transformation eliminates the problem of estimated negative BRDF values. The error matrix $\mathbf{e}_1$ of this approximation can be written as $\mathbf{B}_0 = \mathbf{B}_0' + \mathbf{e}_1$, where $\mathbf{B}_0 = \log(b_{ijkl})$ and $\mathbf{B}_0'$ is the approximation. The approximation is improved by applying recursively the decomposition on the error terms until a satisfactory level of accuracy is obtained; assuming that S is the total number of iterations, the expression of $\mathbf{B}_0$ becomes:

$$\mathbf{B}_0 \approx \mathbf{B}_0' + \mathbf{e}_1' + \mathbf{e}_2' + \ldots + \mathbf{e}_{S-1}', \tag{3.47}$$

where $\mathbf{e}_1 = \mathbf{e}_1' + \mathbf{e}_2$ and $\mathbf{e}_1'$ is the Tucker approximation of $\mathbf{e}_1$, $\mathbf{e}_2$ is the error term of the second, and so on. This non-negative representation allows good compression ratios while being able to represent Fresnel effects and off-specularities but does not satisfy reciprocity and energy conservation. As for the importance sampling, to limit the sampled region for isotropic materials, it has been shown that most of the total variation is due to two components that correspond to univariate functions of θ_h and θ_r; a similar property is observed for the anisotropic material, where the main components are univariate functions of θ_h and ϕ_h.

Pacanowski et al. [PSCS*12] employ a subset of the halfway parameterization [Rus98] to project measured BRDFs on the two-dimensional space (θ_h, θ_d) and approximates the projection by using Rational Functions (RF), since they are able to properly approximate the typical steep changes of specular lobes. An RF r of a finite dimensional vector $\mathbf{x}$ of real variables is defined as:

$$r_{n,m}(\mathbf{x}) = \frac{\sum_{j=0}^{n} p_j b_j(\mathbf{x})}{\sum_{k=0}^{m} q_k b_k(\mathbf{x})}, \tag{3.48}$$

where p_j and q_k are real numbers and $b_j(\mathbf{x})$, $b_k(\mathbf{x})$ are multivariate basic functions, for example multinomials. Given $t + 1$ measured values b_i located at a vector $\mathbf{x}_i$ and contained in the intervals $[\underline{b_i}, \overline{b_i}]$, the data fitting problem can be stated as finding an RF $r_{n,m}(\mathbf{x})$ with the smallest possible $n + m$ to interpolate the $t + 1$ intervals with the additional constraints of non-negativity, monotonicity, and symmetry: $\forall i = 0, \ldots, t$ $\underline{b_i} \le r_{n,m}(\mathbf{x}_i) \le \overline{b_i}$. The widths of the interpolation intervals are chosen in such a way that the renderings are visually satisfactory while keeping the number of coefficients reasonably low; the solution is found by solving a quadratic programming problem. Isotropic BRDF data are approximated with a single RF, called Rational BRDF:

$$f_{r,s}(\mathbf{v_i}, \mathbf{v_r}) \approx f_{\theta_h, \theta_d} \approx r_{n,m}(\theta_h, \theta_d). \tag{3.49}$$

The anisotropic model is based on the observation that for some anisotropic materials, like brushed metals, the variation of the reflected intensity, when the surface is rotated around the normal **n**, consists of a scaling factor applied to an average isotropic lobe:

$$f_{r,s}\left(\mathbf{v_i}, \mathbf{v_r}\right) \approx r^a_{n',m'}(\phi_h) r^i_{n,m}(\theta_h, \theta_d), \tag{3.50}$$

where $r^i_{n,m}(\theta_h, \theta_d)$ is an isotropic Rational BRDF and $r^a_{n',m'}(\phi_h)$ is a scaling factor to model anisotropic variations. The same approximation process applied to BRDFs can be applied to the inverse Cumulative Distribution Function (CDF) and used for importance sampling; the use of RF for BRDFs and CDFs allows obtaining a very small memory footprint.

More recently, tensor representation has been used in [WKB12, WKB14] to represent anisotropic materials with no assumption on the reflectance and scattering behavior, which is particularly useful in the presence of unusual scattering properties. The measured data is fitted to a series of radial basis functions in order to derive a continuous representation from the sparse input 4-D measurements. The incident and reflected hemispheres are projected onto disks and mapped over the unit square; the four dimensions given by the two squares define a rank-4 tensor, subdivided into a tensor tree for fast Monte Carlo sample generation. The tensor tree representation adaptively subdivides sharp peaks of the BRDF in different regions of the distribution, with an additional averaging step between incident and reflected direction to account for Helmoltz reciprocity.

3.4   AVAILABLE RESOURCES FOR BRDF DATA FITTING AND VISUALISATION

Sources of publicly available measured materials are, other than the previously mentioned MERL dataset [Mat03], the UTIA BTFs [HM12] and BRDFs [FV14] datasets, and the Bonn UBO2003 [WGK14] and UBO2014 [WGK14]; as for architectural materials, the *pab BME BSDF gallery* [AB14] provides measured data.

Measured BRDF data usually cannot be used directly for rendering due to the noise in the measurement and needs to be interpolated or fitted to some analytic model. The BRDF analysis library ALTA [BCP*15] provides a set of instructions to perform fitting of the measured data to an analytical form, statistical analysis of data, and offers a wide range of formats to export BRDF models. Additionally it provides functions to handle BRDF models, data to work with BRDF measurements, and filters for fitting algorithms. Dupuy et al. [DHI*15] recently released a $C++$ library to fit a microfacet BRDF to an input material.

Open source graphical user interfaces for BRDF visualization and fitting allow to display and adjust the parameters of several BRDF models. Examples of open source interfaces include BRDFLab [FPBP09] and BRDF Explorer [MHH*12] proposed by Disney (see Figure 3.16 for an example of 2D and 3D visualization of BRDF lobes). These applications handle analytical and measured reflectance models.

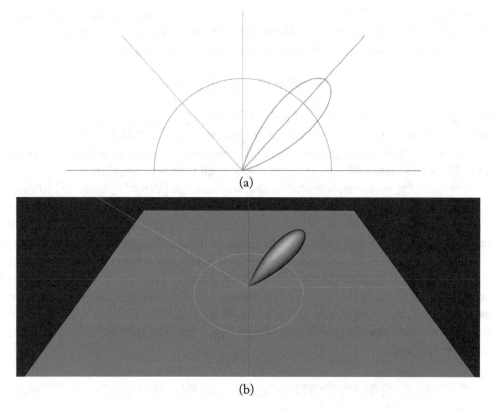

(a)

(b)

Figure 3.16: Cook-Torrance specular lobe. 2D (top) and 3D (bottom) visualization of the lobe in BRDF Explorer [MHH*12].

BRDF Explorer [MHH*12] compares measured material models with existing analytic models and allows interactive adjustment of a few parameters. BRDFLab allows fitting measured data to analytical models and also combining different reflectance models, for example the Lafortune model and Blinn lobes. The software can also perform real-time optimization of the models.

BRDF-Shop [FPBP09] is an interface developed to intuitively design arbitrary but physically plausible BRDFs based on the extended Ward BRDF model. The interface provides control over the surface roughness and sets of brushes to add reflections and highlights. Finally, it is possible to adjust parameters and render the model under simulated lighting; a plug-in for Maya is available to the community.

Currently the libraries of measured materials for rendering applications and the BRDF models included in rendering systems, generally defined as Shaders [Sta99], do not fulfill modern material representation requirements and are computationally expensive. Moreover there is no universal BRDF model that can represent a wide range of materials, since most of them are designed to represent a specific set of phenomena. There are BRDF databases developed for specialized purposes, but to develop new material, a designer usually still needs to start from scratch to display it properly.

CHAPTER 4

Acquisition Setups

Measuring or calculating how a surface interacts with light is a time consuming and expensive procedure, which generates a vast amount of data, but it is important for realistic appearance of a material model. BRDF measurements are not only used in computer graphics to reproduce material reflectance, but also in many other fields such as Computer Vision (e.g., in object recognition applications), Aerospace (e.g., for optimal definition of satellite mirrors' reflectance and scattering properties), Optical Engineering, Remote-Sensing (e.g., land cover classification, correction of view and illumination angle effects, cloud detection, and atmospheric correction), Medical applications (e.g., diagnostics), Art (e.g., 3D printing), and Applied Spectroscopy (e.g., physical condition of a surface).

The setup of a typical measurement device includes a light source to uniformly illuminate a large area of a surface and a detector to measure a small area within the illuminated region [ASMS01]. Various systems with different degrees of accuracy and costs have been constructed to measure reflectance functions, ranging from gonioreflectometers to image-based measurement systems; low cost setups have also been investigated [HP03, FHV15, RLCP]. By dropping the assumption that a material is homogeneous and opaque, many techniques for BRDF measurement can be adapted for more complex reflectance functions (SVBRDFs, BTFs, BSSRDFs). Under certain assumptions, setups used to acquire objects geometry through the classical photometric stereo technique [Woo80], where the point of view is kept constant between successive images while the direction of incident illumination varies, have been successfully used to recover BRDF and SVBRDF of non-Lambertian surfaces [Geo03, GCHS05, HS05, CGS06, ZREB06, AZK08, HLHZ08]. Some of these techniques are limited to materials with a single specular lobe [GCHS05] due to the use of optimization algorithms to recover the parameters for the Ward isotropic BRDF [War92] or require the acquisition of reference objects of known shape and with similar materials as the target [HS05]; to reduce the number of input pictures, have been assumed, bivariate BRDFs [AZK08] or spatial coherence of reflectance, trading spatial for angular resolution [ZREB06]. The taxonomy of the acquisition setups, detailed in the following sections, is reported in Figure 4.1.

To assess the quality of an acquisition setup, it is important to derive a standardized error between the measured appearance model and the original object [GLS04]. The distance metric ΔE is particularly suitable to measure color differences; it is defined in the *CIE XYZ* color space [WS82], a perceptually uniform space that describes the chromatic response of a standard human observer to the lighting stimulus, accounting for the incident spectral power distribution.

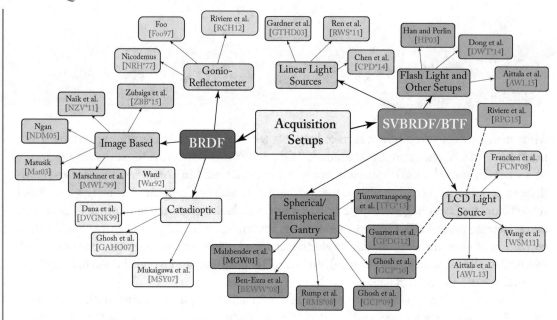

Figure 4.1: **Reflectance measurement setups.**

For digital image sensors, at the heart of image-based systems, the most common color space is *sRGB*, which due to the characteristics of the Human Visual System is often prone to inaccuracies [Fai05]. Acquired RGB values could be translated into the *CIE XYZ* and post-processed for white balancing [WEV02], although metamerism (i.e., spectra that appear identical to a human observer under a certain light) would still represent a source of errors. A more robust solution would make use of a carefully characterized acquisition device (e.g., a DSLR camera) to obtain either a relative [KK08] or an absolute colorimetric estimation of the scene in cd/m^2 [GBS14].

Numerical simulation [CMS87, WAT92, HK93, APS00, DWMG15] represents, for some complex materials, a possible alternative to a measurement device. The material appearance is described by the result of the simulation of the light interaction with the surface (and sub-surface) structure. Given a geometry that can be ray-traced, Westin et al. in their seminal work [WAT92] describe a method to simulate scattering hierarchically by using the result of the simulation at a scale to generate the BRDF for a larger scale.

The measurements are always made with some uncertainty caused by technological limitations of each measurement system component, hence estimating the BRDF of a sample material with low uncertainty would be expensive.

4.1 GONIOREFLECTOMETERS

The gonioreflectometer measures the spectral reflectance of surfaces; it covers specular and diffuse reflectance depending on the settings of the device. The construction of the device was described by Nicodemus and used in the experimental development of the reflection models by Torrance and Sparrow [TS67], Blinn [Bli77], He et al. [HTSG91], and many others.

The setup proposed by Hsia and Richmond [HR76] consists of a laser beam, an aluminium sample holder painted with matte black paint and mounted on an arm attached to a turntable that rotates around the vertical axis. The turntable is placed in front of a detector, which captures data about light reflected from the sample. Two averaging spheres, with the inside part coated with barium sulfate, are used to measure the incident light.

Foo [Foo97] designed a three-axis automated gonioreflectometer with two degrees of freedom (see Figure 4.2). The measuring system, aimed to isotropic BRDFs, consists of a light source moving around a sample, a stationary detector, and a folding mirror. The system can also measure the reflection toward grazing angles (up to 86°) and allows high-dynamic range measurements, making it considerably accurate. Li et al. [LFTW06] described a similar setup.

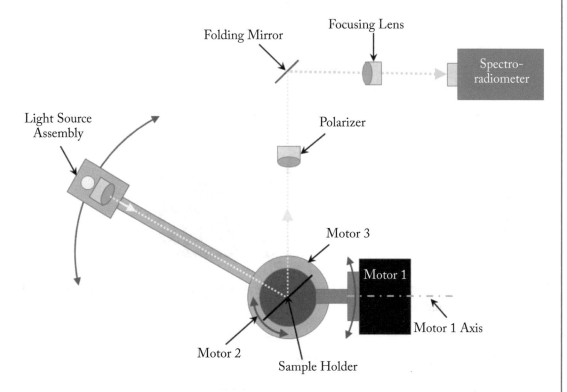

Figure 4.2: The three-axis automated gonioreflectometer proposed by Foo [Foo97].

Riviere et al. [RCH12] proposed an in-plane multispectral polarized reflectometer. The lighting system makes use of three linearly polarized laser sources, coupled with a polarized detection system that identifies the polarizer's axes using the Fresnel Equation (Figure 4.3). It allows measurements at 0° lighting and is fully calibrated for polarized and multispectral in-plane BRDF measurements. Polarized measurements are used to distinguish the different scattering processes in BRDF directional components. Since the acquired data is limited to in-plane measurements, the full BRDF is estimated by fitting the data to the Li-Torrance model [LT05].

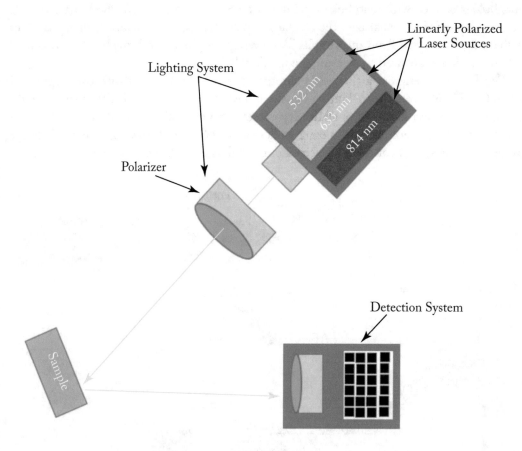

Figure 4.3: In-plane multispectral polarized reflectometer proposed by Riviere et al. [RCH12] (schematic representation).

4.2 IMAGE-BASED MEASUREMENT

Image-based BRDF measurement setups aim to increase the speed of BRDF measurements and reduce the cost by making use of a series of photographs of a surface and typically requiring only general-purpose equipment. These photographs are used to estimate the light reflected from various surface orientations; to measure the wavelength spectrum of the BRDF, more pictures, and hence, time, are required [MD98].

Marschner et al. [MWL*99] presented a rapid and accurate isotropic BRDF measurement setup for a broad range of homogeneous materials, including human skin. It can achieve high resolution and accuracy over a large range of illumination and reflection directions. A handheld digital camera, equipped with a standard CCD sensor with RGB color filter array, is used to measure the BRDF. The camera needs to be characterized in terms of Optoelectronic Conversion Function (OECF) in order to properly estimate the reflected radiance and the irradiance due to the source. An industrial electronic flash light is used as a light source (see Figure 4.4 for a schematic representation). This setup allows measuring surfaces with simple analytical shapes, for example, spherical and cylindrical objects; a 3D scanner is required for more complex irregular shapes. The camera moves from near the light source, to measure near retro-reflections, and to opposite the light source, in order to measure grazing-angle reflection. Some additional photographs are taken by the secondary camera in order to infer the location of the light source (by means of some machine-readable targets) and its intensity other than from camera and sample pose. About 30 images from different positions are required to cover the three-dimensional BRDF domain, and a typical measurement session takes about 30 min. Bundle adjustment is used to estimate the relationship between the geometry of the sample and the positions of the camera, light source, and sample; thanks to the estimated geometry of the scene, each pixel in the images is used to derive one sample in the BRDF domain.

A few years later, Matusik et al. [MPBM03a] used a measurement setup similar to Marschner et al. [MWL*99] to capture isotropic BRDFs, stored in a tabular form and interpolated linearly using a small number of basis functions to cover the entire space [OKBG08]. Matusik's data-driven method is described in Section 3.3.1. The main difference with the acquisition setup proposed by Marschner et al. is the use of a turntable, with an arm attached to it holding the light source: this allows removing the secondary camera, since the position of the light source is known and accurately controllable (Figure 4.5). This measurement setup has been used to capture a dataset of 100 isotropic materials, which is to date still widely used in computer graphics and other fields; Figure 4.6 reports some renderings of the BRDFs included in the data set.

Ngan et al. [NDM05] presented an anisotropic BRDF acquisition setup for flat and flexible samples wrapped around a cylinder of known size; the cylinder can be tilted by means of a precision motor in order to account for the missing degree of freedom with respect to a sphere. To deal with the anisotropy, strips of the samples at different orientations are wrapped around the cylinder (see Figure 4.7). A light source rotates around the cylinder while the target is captured by a fixed camera, and for each light and target position a set of eight pictures with different exposures is taken to form an HDR image, enabling the capture of the full 4D BRDF. The sampling density of the light and the cylinder tilting can be adjusted to increase the resolution of

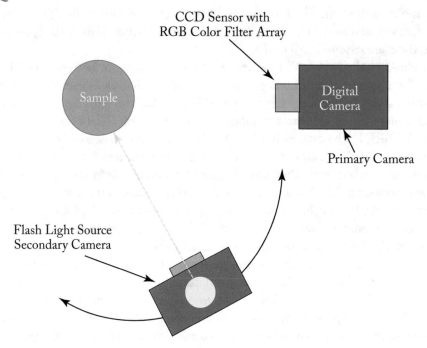

Figure 4.4: The image-based acquisition setup for isotropic BRDFs proposed by Marschner et al. [MWL*99].

the measured BRDF. The number of material strips that can be wrapped around the cylinder is clearly limited and represents the main limitation in the resolution. The reflectance acquisition setup proposed by Naik et al. [NZV*11] exploits space-time images captured by a time-of-flight camera. Two different setups are described, both based on indirect viewing with three-bounce scattering and making use of two known Lambertian materials, respectively the source S and the receiver R, while P is the patch to measure. In the first setup, the laser illuminates S, and the camera views R, thus measuring P indirectly. As for the second configuration, it is based on an around-the-corner viewing in which P is not directly visible to the camera, whereas S and R are the same surface. The light is multiplexed along different transport paths and some of them might have the same length, hence the light can arrive along multiple paths at the same point at the same time. For this reason the measurements of the material need to be decoded by solving a sparse under-determined system; the system is solved by recovering the parameters with the Ashikhmin-Premoze model [AP07] (see Section 3.1.2), using the halfway vector parameterization. When the multiplexing does not cause ambiguities, in order to measure the parameters of a material, it is enough to analyze the streak images to find the specular peak. This setup enables to taking many BRDF measurements simultaneously, but it requires an ultra-fast camera; moreover it suffers from a low signal-to-noise ratio due to the multiple bounces; the size of patches

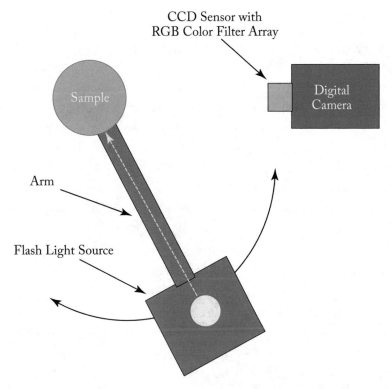

Figure 4.5: The image-based acquisition setup proposed by Matusik et al. [MPBM03a].

and the maximum sharpness of the reflectance function are limited by the hardware, and the range of measurable incoming and outgoing directions is limited by the geometry of the setup.

4.3 CATADIOPTRIC MEASUREMENT SETUPS

Catadioptric optical systems measure both reflected and refracted light, thus reducing aberrations. Setups in this category usually do not make use of moving parts and tend to be efficient image-based BRDF acquisition devices.

The imaging gonioreflectometer described by Ward [War92] is based on a half-silvered hemisphere and a fisheye lens. It captures the entire hemisphere of reflected and refracted directions at the same time and allows measuring anisotropic surfaces by repeating the measurement process under various orientations (Figure 4.8). This device cannot measure sharp specular peaks or take measurements at high grazing angles.

The Dana et al. [DVGNK99] measurement device consists of a robotic arm that holds and rotates a sample, a halogen bulb with a Fresnel lens, and a video camera. The light position is kept fixed, while the camera moves to record measurements from seven different locations

Figure 4.6: Subset of 64 materials, out of the 100 in the MERL-MIT BRDF database by Matusik et al. [MPBM03a].

(Figure 4.9). To enable measurements of anisotropic materials, the sample is rotated around the z-axis and the procedure is repeated. This system was designed for use in computer graphics, and like Ward, includes reflection and refraction capture; however, there are issues with noise since the sample patches are too large to measure fine-scale texture variations.

Mukaigawa et al. [MSY07] built a measurement system for anisotropic BRDFs that uses a projector as the light source placed at the focal point of an ellipsoidal mirror, a camera, and a beam splitter, since the camera and the projector cannot be located at the same position (Figure 4.10). The number of acquired images depends on the desired sampling of lighting and

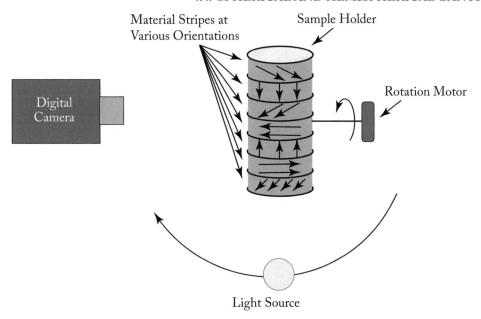

Figure 4.7: Anisotropic BRDF acquisition setup proposed by Ngan et al. [NDM05].

viewing directions, which needs to be estimated based on the accuracy required. The acquired data are then fitted to the Ward anisotropic reflection model (Section 3.1.2).

Ghosh et al. [GAHO07, GHAO10] developed a measurement device that consists of a camera focusing on a zone of reflected directions, a light source with a beam splitter, a mirrored dome, and mirrored parabola (Figure 4.11). The focus of the illumination beam is on the mirrored components that the beam reflects back to its origin. This setup allows BRDF measurement over a continuous region with a specially designed orthonormal zonal basis function illumination, allowing a very rapid BRDF acquisition and a better signal-to-noise ratio compared to point-sampling the incident directions, as in [MSY07]. Measured data is then projected into a spherical harmonics basis or fitted to an analytical reflection model.

4.4   SPHERICAL AND HEMISPHERICAL GANTRY

Malzbender et al. [MGW01] built a hemispherical device suitable to measure almost-flat samples (Figure 4.12). A camera is placed in the apex of the device, while the flat samples are placed on the floor and illuminated by a single light source at a time out of the 50 strobe lights available. The acquired data are represented by Polynomial Texture Maps (PTM), in which for each texel

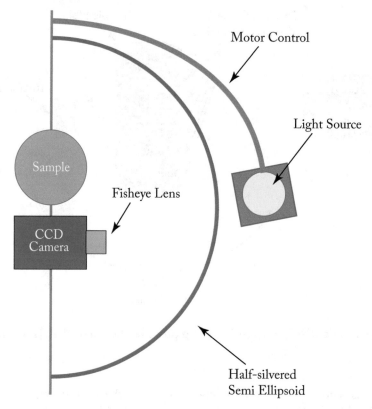

Figure 4.8: Acquisition setup proposed by Ward [War92].

the coefficients of the following polynomial are fitted to the data and stored as a map:

$$L(u,v;l_u,l_v) = a_0(u,v)\,l^2{}_u + a_1(u,v)\,l^2{}_v + \\ + a_2(u,v)\,l_u l_v + a_3(u,v)\,l_u + a_4(u,v)\,l_v + a_5(u,v),$$

(4.1)

where L is the surface luminance at (u,v); the local coordinates of the texture and (l_u,l_v) are the projection of the normalized light vector at that coordinate. PTMs facilitate good quality rendering, in particular for almost diffuse samples.

A hemispherical device for anisotropic BRDF measurement, which does not require any moving parts or cameras, was presented by Ben-Ezra et al. [BEWW*08]. This work demonstrates that an accurate radiometric and geometric calibration allows using LEDs both as light sources and as detectors (Figure 4.13) with fast acquisition times. In their implementation 84 LEDs are used, pointing toward the center of the hemisphere. During the acquisition each LED is switched on, in turn acting as an emitter, while all others measure the light reflected from the sample. The signal-to-noise ratio tends to be low, but it can be increased by means of

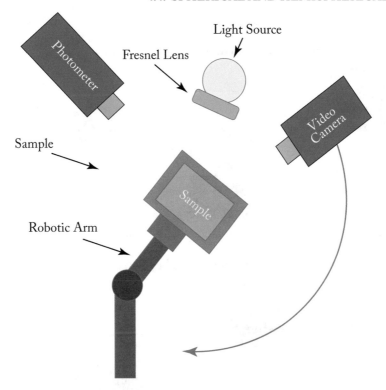

Figure 4.9: Acquisition setup proposed by Dana et al. [DVGNK99].

multiplexed illumination; the use of different colors for the LEDs allows acquiring multispectral data. Since an LED cannot be used at the same time as an emitter and detector, this setup cannot be used to measure retro-reflection and offers a lower resolution compared to camera-based setups.

The measurement device presented by Rump et al. [RMS*08] consists of a hemispherical gantry with 151 cameras uniformly distributed; the cameras flashes are used as light sources, and for each flash all the cameras take a picture of the subject, giving a total of $151 \times 151 = 22,801$ pictures, which can be increased by taking HDR sequences. The gantry is capable of supporting projectors in order to use structured light on the subject.

Ghosh et al. [GCP*09] proposed three different setups to estimate spatially varying BRDFs for both isotropic and anisotropic materials, using up to nine polarized second-order spherical gradient illumination patterns. For specular reflections, specular albedo, reflection vector, and specular roughness can be directly estimated from the 0th, 1st [MHP*07], and 2nd order [GCP*09] statistics respectively. The first setup, suitable for roughly specular objects of any shape, is based on an LED sphere with 150 controllable lights linearly polarized, with the subject placed at the center of the sphere. The second setup is suitable for flat objects and uses

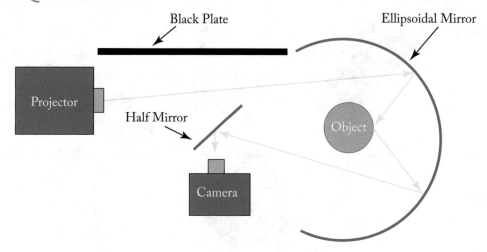

Figure 4.10: Catadioptric acquisition setup proposed by Mukaigawa et al. [MSY07].

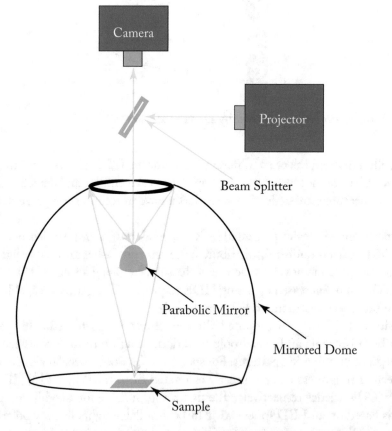

Figure 4.11: Reflectance acquisition setup by Ghosh et al. [GAHO07, GHAO10].

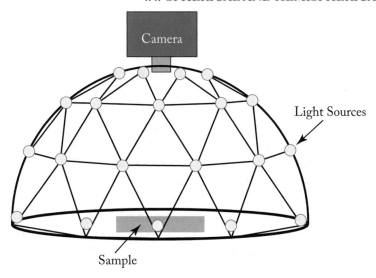

Figure 4.12: Schematic representation of the hemispherical device proposed by Malzbender et al. [MGW01].

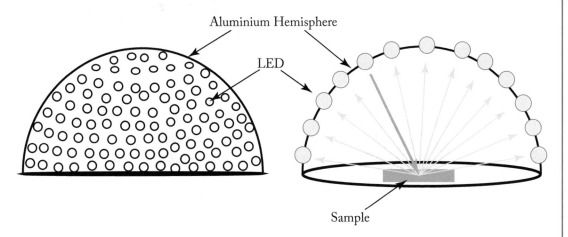

Figure 4.13: Schematic representation of the LED-only BRDF measurement device proposed by Ben-Ezra et al. [BEWW*08].

as the light source an LCD monitor, placed very close to the subject, which clearly offers a smaller coverage of incident direction but with a higher resolution than the LED sphere. The third setup makes use of a roughly specular hemisphere that reflects the light emitted by a projector on the subject placed at the center of the hemisphere, thus allowing a dense sampling; the camera observes the subject from the apex of the hemisphere.

The analysis of the Stokes reflectance field of circularly polarized spherical illumination has been exploited by Ghosh et al. [GCP*10] to estimate the specular and diffuse albedo, index of refraction, and specular roughness for isotropic SVBRDFs, assuming known surface orientation. Three different setups are used to demonstrate the technique, similar to the ones described in [GCP*09] but with the light sources covered with right circular polarizers. Four pictures of the subject are required to measure the Stokes field, three of them with differently oriented linear polarizers in front of the camera and one with a circular polarizer.

The same framework based on the analysis of the Stokes reflectance field has been further exploited by Guarnera et al. [GPDG12] and is extended to also cover unpolarized illumination to obtain a per-pixel estimate of the surface normal from the same input data as in [GCP*10]. The proposed setup makes use of an LED sphere with 346 controllable lights unpolarized/circularly polarized; the surface normals estimation is also demonstrated with uncontrolled outdoors measurement under overcast sky (hence unpolarized lighting) by capturing a reference dielectric sphere in the same environment as the object, which can be of arbitrary shape (see Figure 4.14).

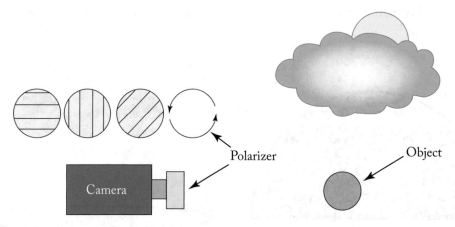

Figure 4.14: Surface normal estimation under uncontrolled lighting (outdoor, overcast sky), described in [GPDG12].

Tunwattanapong et al. [TFG*13] proposed a spinning spherical reflectance acquisition apparatus. A spinning semi-circular arc of 1 m diameter, with 105 LED focused toward the center, rotates about the vertical axis at 1 rpm, sweeping out continuous spherical harmonic illumination conditions (Figure 4.15). Thanks to the spherical harmonic illumination, 44 pictures, taken by a five-camera array, are enough to estimate anisotropic SVBRDFs and the 3D geometry of very specular or diffuse objects. The technique can be considered as a further generalization of the approach by Ghosh et al. [GCP*09], since it makes use of higher-order spherical harmonic illumination (up to 5th order), which allows obtaining diffuse/specular separation without relying on polarization.

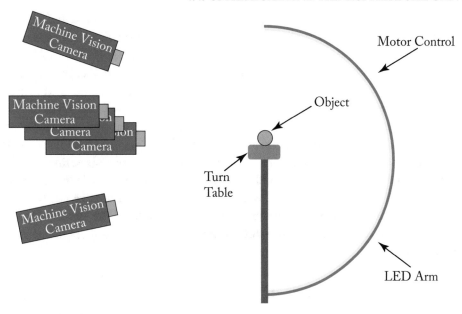

Figure 4.15: Acquisition setup by Tunwattanapong et al. [TFG*13].

Gardner et al. [GTHD03] built a low cost linear light source apparatus to capture flat samples making use of a fixed camera for imaging and a structured light diode. The light source is a 50 cm long neon tube, which is translated horizontally over the surface of the subject and moved in sync with the camera acquisitions. The reflectance model used to fit the measured data is the isotropic model by Ward [War92], given the camera and light source positions at each frame. The laser projects a laser stripe, which is deformed by surface variations and used in order to recover the geometry, together with two scans of the light source, in a diagonal direction. A cabin light box, with two diffused cathode tubes, is used as a sample holder and to project an even diffuse white light on the surface and allows measurement of the transmitted and reflected light. Overall, the system allows recovery of the diffuse and specular colors, specular roughness, surface normals, and per-pixel translucency for isotropic samples.

In Ren et al. [RWS*11] a handheld linear light source device, together with a BRDF chart, is employed to obtain spatially varying isotropic BRDFs from a video taken with a mobile phone in LDR. The BRDF chart consists of 24 square flat tiles with known BRDFs. The tiles are made of specular materials, except one that is a diffuse standard for camera calibration (exposure and white balance). The light source is a 40 cm florescent tube, slowly moved by hand over the surface and the chart, which needs to be placed alongside. This approach requires solving a number of issues, since the camera and the light source need to be placed close to the sample and the light is moved manually. Consequently, the camera and light position are unknown, as well as the SVBRDF of the sample. Saturated values from LDR acquisition are repaired using the values in the neighborhood, and the reflectance responses are normalized and hence aligned by a dynamic time-warping algorithm. Aligned samples are then used for BRDF reconstruction.

Chen et al. [CDP*14] present a similar setup to Gardner et al. [GTHD03], scanning a linear light source over a flat sample but with the significant advantage of capturing anisotropic surface reflectance. The basic assumption is that a microfacet model can be used to model the anisotropic surface reflectance. To observe the specular reflection, they modulate the illumination along the light source by means of a transparent mask. They propose two different setups that differ in form factor and employ the same 35 cm CCFL lamp and DSLR camera. The desktop form factor scanner scans a linear light source over the sample, observing the SVBRDF by means of the camera; as for the handheld form factor scanner, the sample moves with respect to the camera and the linear light source, which instead has a fixed relative position. Finally a cylindrical lens is employed to capture in a single picture a scanline of the sample. One constant lighting pattern, together with two phase-shifted sinusoidal patterns, suffices to reconstruct the surface reflectance.

4.5   LCD LIGHT SOURCE

Francken et al. [FCM*08] make use of commodity hardware such as an LCD display and an SLR camera to recover detailed normal maps of specular objects, based on the observation that the normal of a specular pixel is the halfway vector between the light direction and the view direction. To identify the light direction among n different light sources, they make use of gray code lighting patterns by taking $O(\log_2 n)$ pictures. The accuracy of the estimated normal map depends on the number of sampled light sources.

In Aittala et al. [AWL13] a low cost capture setup for SVBRDFs is presented with a similar setup as Francken et al. Their work relies on the design of the image formation model and uses a Fourier basis for the measurements. Isotropic BRDFs are reconstructed through Bayesian inference since the model is analytically integrable.

The capture setup by Wang et al. [WSM11] consists of a vision camera and a regular LCD used as an area light source. It allows rapid measurement of a stationary, isotropic, glossy and bumpy surface, describing its appearance with a dual-level model, which consists of the specular and diffuse relative albedos, two surface roughness parameters, and a 1D power spectrum over frequencies for visible surface bumps. Two images are required for calibration, since the LCD radiance is dependent on the viewing angle. The first one captures the surface reflection while displaying a constant gray image on the LCD, and the second one is taken by attaching a mirror to the surface. To establish the pose of the surface with respect to the camera, a target is placed on the surface. At the microscale, the reflectance is characterized with the Cook-Torrance model and the distribution D is assumed to be Gaussian, where the standard deviation represents the roughness; similarly at the mesoscale level, roughness is approximated in terms of the standard deviation. The effect of the roughness at the microscale is assumed to be a blurring of perfect mirror reflections, whereas at the mesoscale it determines a permutation of the pixels. The surface is illuminated with a half-black, half-white image with a vertical edge, and the overall roughness is estimated by fitting a Gaussian filter that blurs the step-edge image to produce the observed one. To separate the roughness for the two different scales, all pixels are sorted by intensity and reshaped back in column-major order, thus removing the permutation induced by the mesoscale roughness; the slope of the segment obtained by averaging over the rows of the sorted image is used to estimate the microscale roughness. This approach can produce visually plausible results for highly glossy manmade indoor surfaces, including some paints, metals, and plastics.

Riviere et al. [RPG15] propose a mobile reflectometry solution based on a mobile device's LCD panel as an extended illumination source, statically mounted at a distance of 45 cm above an isotropic planar material sample, at normal incidence, in a dimly lit room. The linear polarization of the LCD panel is exploited for diffuse/specular separation by taking two pictures of the sample with a differently oriented plastic sheet linear polarizer in front of the device camera. Albedo, surface normals, and specular roughness are estimated by illuminating the sample with the same lighting patterns described in [GCP*09]. Due to the limited size of the LCD panel and the position of the front camera, this setup can only acquire a 5 cm × 5 cm area of the sample; for larger samples, an appearance transfer approach that relies on additional measurements under natural illumination is used.

4.6   FLASH ILLUMINATION AND OTHER CAPTURE SETUPS

Backscattering data can be used to extract an appropriate distribution for microfacets BRDF models [AP07]. Based on this observation, mobile devices equipped with a flash light, typically near the back camera, represent near-coaxial setups particularly useful to capture the backscatter surface reflectance to be fitted in a microfacets BRDF model [RPG15].

Riviere et al. [RPG15] mobile flash-based acquisition setup estimates the diffuse and specular albedo, surface normal, and specular roughness of a planar material sample, with spatially varying isotropic surface reflectance. The back camera and flash light of a mobile device are used for a handheld acquisition of a video in a dimly lit room, capturing data of the sample from several directions over the upper hemisphere (Figure 4.16). For reflectance calibration, the diffuse gray squares of an X-Rite ColorChecker are used. The top view of the sample at normal incidence is used as a reference to register the other frames. To estimate the lighting and view directions, the magnetometer/accelerometer sensors or 3D tracking can be used. The surface normal of each point is computed as the weighted average of the brightest reflection direction; the diffuse albedo is estimated as the trimmed median of the measured intensities, whereas the specular albedo is estimated from the hemispherical integral of the diffuse subtracted measurements. The specular roughness is obtained by fitting the observed backscattering profile to the [WMLT07] model (see Section 3.2.1). Some blurring in the reflectance maps can be introduced by misalignments and motion blur. The limited number of lighting directions suggests use only for rough specular materials.

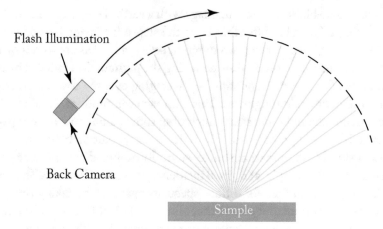

Figure 4.16: Mobile flash-based reflectance acquisition setup by Riviere et al. [RPG15], based on backscattering measurements with flash illumination.

Aittala et al. [AWL15] mobile measurement setup for stationary materials consists of a single mobile device with onboard flash light (Figure 4.17). Given a flash-no-flash image pair of a textured material of known characteristic size, a multistage reconstruction pipeline allows capturing the full anisotropic SVBRDF. The input images are registered through a homography, computed from manually specified points of correspondence. The flash image provides an approximate retro-reflective measurement for each pixel that combines the effect of surface normal and BRDF, whereas the other image is used as a guide to identify points on the surface with similar local reflectance. Since there is only one observation per pixel, it is assumed that multiple points on the surface share the same reflectance properties and can be identified under ambient lighting to be combined together. The input is organized into regular tiles approximately the same size of the repeating texture pattern, assumed to contain a random rearrangement of the same BRDF values. A mastertile is selected for relighting, and lumitexels (i.e., data structures to store the geometric and photometric data of one point [LKG*03]) are obtained for it. The lumitextels are regularized using a preliminary SVBRDF fit and augmented by transferring high-frequency detail from similarly lit tiles to reduce blurring. The augmented lumitexels are used in a non-linear optimizer to fit an analytic SVBRDF model and the solution is finally reverse-propagated to the full image. This setup limits the input to the retro-reflective slice of the BRDF, hence the Fresnel effect; shadowing and masking are assumed to have typical behavior and modeled with the BRDF model A [BLPW14] (see Section 3.1.2). The camera field of view represents an upper limit on the width of the specular lobes that can be observed.

The idea that the variation of the reflectance over a target forms a low-dimensional manifold is exploited by Dong et al. [DWT*10] and describes a two-pass method to accelerate complex reflectance capture, useful for both isotropic and anisotropic flat samples. During the first

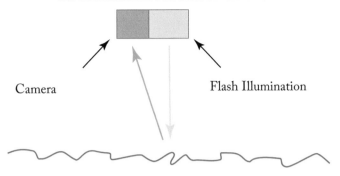

Figure 4.17: Geometry of the imaging setup proposed by Aittala et al. [AWL15].

phase a set of high-resolution representative single-point BRDFs is captured using a handheld device that scans over the sample. The device consists of a pair of condenser lenses, a pinhole, and a camera, aligned along the same optical axis by means of a lens tube. Six high-brightness LEDs are used as light sources, with one light at the top and the remaining at the sides. The pinhole is placed at the focal plane of the ocular condenser lens so the camera can image the light field of a single point on the surface, while the sample is placed at the focal plane of the field condenser lens. For each light, a pair of 320×240 pixel pictures is taken, with different exposures in order to obtain a 240×240 HDR image, used for local reconstruction of BRDFs by convex linear combination in a small neighborhood. The second phase captures a set of reflectance measurements densely over the surface by means of a fixed DSLR camera and a handheld light source, about 1.5 m away from the sample and moved in a 2D plane opposite the sample itself; a mirrored ball is used to sample the incident lighting. Up to 200 pictures are acquired and used to map the manifold derived from the first phase over the sample surface.

The measurement device presented in [HP03] is based on the principle of the kaleidoscope and consists of a tapered tube whose inner walls are lined with front-surface mirrors. A single camera captures the kaleidoscopic image, in which the sub-images represent the same sample seen simultaneously from many different viewpoints. The sample is illuminated by a DLP projector, which shares the optical path with the camera by means of a 45° beam splitter. The properties of the sample are measured through a sequence of pictures with different illumination images, which illuminate the sample from a known range of incoming directions due to the unique sequence of reflections from the kaleidoscopic walls. The advantages of this setting, suitable for BTFs and BSSRDFs, are the absence of moving parts, which enables quick measurements and guarantee perfect registration of the measurements and the low cost; radiometric and geometric calibration need to be performed only once.

CHAPTER 5

Conclusion

5.1 CONCLUSION

We described the class or reflectance models, with a particular focus on surface scattering models (BRDF) their representations, and acquisition setups. Each model is limited by the underlying assumptions and the set of parameters to implement it, which result in the ability of the model to represent a specific material group. Even generalized models cannot cover a broad range of materials or all the variations of a material within one group.

Currently material modeling often involves a great deal of manual effort, although available solutions range from completely manual creation of a material to fully automated acquired material, the latter rarely suitable for a straightforward application in rendering. The broad range of material models and complexity of the parameters requires from an artist a deep understanding of the underlying representation and material's micro/macrostructure.

There is no straightforward pipeline to measure and represent BRDF data, and not all the materials can be captured with existing reflectance measurement setups due to their peculiarities (some smart materials for instance), hence it might happen that a material needs to be rendered before a robust measurement and representation are available. These materials cannot inherit from the base materials available in the rendering tool, and existing model assets cannot be reused. This raises the issue that a large number of materials still needs to be captured and represented for the computer graphics community.

Some setups require samples of a specific size and shape, with many of them limited to planar, simple spherical, and cylindrical samples; unfortunately it is not always possible to build such samples. Some layered materials, like varnished wood, metallic, or iridescent car paint pose additional issues both on the acquisition and representation side, since their structure contributes to obtaining complex reflectance properties. Although attempts to generalize reflectance models have been made by researchers, there is still no up-to-date universal material representation model that can fulfill such criteria and make it possible to standardize material representation. The development of the appropriate resources continues, and a physically accurate, consistent, and intuitive material representation would be beneficial for computer graphics and related fields.

Bibliography

[AB14] Apian-Bennewitz P. Building material examples (BME) BRDF and BSDF database. http://www.pab.eu/gonio-photometer/demodata/bme/, 2014. 45

[AMHH08] Akenine-Möller T., Haines E., and Hoffman N. *Real-time Rendering*, 3rd ed., A. K. Peters, Ltd., Natick, MA, 2008. 15, 38

[AP07] Ashikhmin M. and Premoze S. Distribution-based BRDFs, Tech. rep., 2007. 28, 54, 65

[APS00] Ashikmin M., Premože S., and Shirley P. A microfacet-based BRDF generator. In *Proc. of the 27th Annual Conference on Computer Graphics and Interactive Techniques (SIGGRAPH)*, pp. 65–74, ACM Press/Addison-Wesley Publishing Co., New York, 2000. DOI: 10.1145/344779.344814. 50

[Arv95] Arvo J. Applications of irradiance tensors to the simulation of non-lambertian phenomena. In *Proc. of the 22nd Annual Conference on Computer Graphics and Interactive Techniques (SIGGRAPH)*, pp. 335–342, ACM, New York, 1995. DOI: 10.1145/218380.218467. 21

[AS00] Ashikhmin M. and Shirley P. An anisotropic phong BRDF model. *Journal of Graphic Tools 5*, 2, pp. 25–32, February 2000. DOI: 10.1080/10867651.2000.10487522. 26, 28, 43

[ASMS01] Ashikhmin M., Shirley P., Marschner S. and Stam J. State of the art in modeling and measuring of surface reflection. In *ACM SIGGRAPH Courses*, p. 1, 2001. 8, 49

[AWL13] Aittala M., Weyrich T., and Lehtinen J. Practical SVBRDF capture in the frequency domain. *ACM Transactions on Graphics 32*, 4, pp. 110:1–110:12, July 2013. http://doi.acm.org/10.1145/2461912.2461978 DOI: 10.1145/2461912.2461978. 64

[AWL15] Aittala M., Weyrich T., and Lehtinen J. Two-shot SVBRDF capture for stationary materials. *ACM Transactions on Graphics 34*, 4, pp. 110:1–110:13, July 2015. http://doi.acm.org/10.1145/2766967 DOI: 10.1145/2766967. 8, 66, 67

[AZK08]   Alldrin N., Zickler T., and Kriegman D.   Photometric stereo with non-parametric and spatially-varying reflectance. In *Computer Vision and Pattern Recognition, (CVPR). IEEE Conference on*, pp. 1–8, 2008. DOI: 10.1109/cvpr.2008.4587656. 49

[BBP15]   Barla P., Belcour L., and Pacanowski R.   In praise of an alternative BRDF parametrization. In *Workshop on Material Appearance Modeling. Proc. of the Workshop on Material Appearance Modeling*, Darmstadt, Germany, June 2015. 10

[BCP*15]   Belcour L., Courtes L., Pacanowski R., et al. ALTA: A BRDF Analysis Library. http://alta.gforge.inria.fr/, 2013–2015. 45

[BEWW*08]   Ben-Ezra M., Wang J., Wilburn B., Li X., and Ma L.   An led-only BRDF measurement device. In *Computer Vision and Pattern Recognition, (CVPR). IEEE Conference on*, pp. 1–8, June 2008. http://dx.doi.org/10.1109/CVPR.2008.4587766 DOI: 10.1109/cvpr.2008.4587766. 58, 61

[Bli77]   Blinn J. F.   Models of light reflection for computer synthesized pictures. *SIGGRAPH Computer Graphics 11*, 2, pp. 192–198, July 1977. DOI: 10.1145/965141.563893. 21, 25, 51

[BLPW14]   Brady A., Lawrence J., Peers P., and Weimer W. GENBRDF: Discovering new analytic BRDFs with genetic programming. *ACM Transactions on Graphics 33*, 4, pp. 114:1–114:11, July 2014. DOI: 10.1145/2601097.2601193. 23, 66

[BÖK11]   Bilgili A., Öztürk A., and Kurt M.   A general BRDF representation based on tensor decomposition. *Computer Graphics Forum 30*, 8, pp. 2427–2439, 2011. DOI: 10.1111/j.1467-8659.2011.02072.x. 43

[Bra02]   Brand M. Charting a manifold. In *Advances in Neural Information Processing Systems 15*, pp. 961–968, MIT Press, 2002. 40

[BS64]   Beckmann, P. and Spizzichino, A. *The scattering of electromagnetic waves from rough surface*, Pergamon Press, 1963. 29

[BSH12]   Bagher M. M., Soler C., and Holzschuch N.   Accurate fitting of measured reflectances using a shifted gamma micro-facet distribution.   In *Computer Graphics Forum*, vol. 31, pp. 1509–1518, Wiley Online Library, 2012. DOI: 10.1111/j.1467-8659.2012.03147.x. 33, 35

[CBP04]   Claustres L., Boucher Y., and Paulin M. Wavelet-based modeling of spectral bidirectional reflectance distribution function data. *Optical Engineering 43*, 10, pp. 2327–2339, 2004. DOI: 10.1117/1.1789138. 38

[CDP*14] Chen G., Dong Y., Peers P., Zhang J., and Tong X. Reflectance scanning: Estimating shading frame and BRDF with generalized linear light sources. *ACM Transactions on Graphics 33*, 4, pp. 117:1–117:11, July 2014. http://doi.acm.org/10.1145/2601097.2601180 DOI: 10.1145/2601097.2601180. 64

[CGS06] Chen T., Goesele M., and Seidel H.-P. Mesostructure from specularity. In *Computer Vision and Pattern Recognition, IEEE Computer Society Conference on*, vol. 2, pp. 1825–1832, 2006. DOI: 10.1109/cvpr.2006.182. 49

[CJAMJ05] Clarberg P., Jarosz W., Akenine-Möller T., and Jensen H. W. Wavelet importance sampling: Efficiently evaluating products of complex functions. In *ACM Transactions on Graphics (TOG)*, vol. 24, pp. 1166–1175, 2005. DOI: 10.1145/1073204.1073328. 17

[CMS87] Cabral B., Max N., and Springmeyer R. Bidirectional reflection functions from surface bump maps. *SIGGRAPH Computer Graphics 21*, 4, pp. 273–281, August 1987. DOI: 10.1145/37402.37434. 50

[CP85] Clarke F. and Parry D. Helmholtz reciprocity: Its validity and application to reflectometry. *Lighting Research and Technology 17*, 1, pp. 1–11, 1985. DOI: 10.1177/14771535850170010301. 13

[CPB03] Claustres L., Paulin M., and Boucher Y. BRDF measurement modelling using wavelets for efficient path tracing. *Computer Graphics Forum*, pp. 701–716, 2003. DOI: 10.1111/j.1467-8659..00718.x. 38

[CT82] Cook R. L. and Torrance K. E. A reflectance model for computer graphics. *ACM Transactions on Graphics (TOG) 1*, 1, pp. 7–24, 1982. DOI: 10.1145/357290.357293. 28, 29, 30, 31, 32, 34, 35, 36

[CT91] Church E. L. and Takacs P. Z. Optimal estimation of finish parameters. In *Proc. SPIE*, vol. 1530, pp. 71–85, 1991. DOI: 10.1117/12.50498. 35

[CTL90] Church E. L., Takacs P. Z., and Leonard T. A. The prediction of BRDFs from surface profile measurements. In *Proc. SPIE*, vol. 1165, pp. 136–150, 1990. DOI: 10.1117/12.962842. 35

[DHI*15] Dupuy J., Heitz E., Iehl J.-C., Poulin P., and Ostromoukhov V. Extracting microfacet-based BRDF parameters from arbitrary materials with power iterations. *Computer Graphics Forum 34*, 4, pp. 21–30, 2015. DOI: 10.1111/cgf.12675. 37, 45

[DI11] D'Eon E. and Irving G. A quantized-diffusion model for rendering translucent materials. *ACM Transactions on Graphics 30*, 4, pp. 56:1–56:14, July 2011. DOI: 10.1145/2010324.1964951. 9

[DR05]   Dorsey J. and Rushmeier H.   Digital modeling of the appearance of materials. In *ACM SIGGRAPH Courses*, p. 1, 2005. DOI: 10.1145/1198555.1198694. 5

[DRS07] Dorsey J., Rushmeier H., and Sillion F. *Digital Modeling of Material Appearance*, Morgan Kaufmann, 2007. DOI: 10.1016/B978-0-12-221181-2.X5001-0. 14

[DS03]   Dachsbacher C. and Stamminger M. Translucent shadow maps. In *Proc. of the 14th Eurographics Workshop on Rendering (EGRW)*, pp. 197–201, Eurographics Association, Aire-la-Ville, Switzerland, 2003. 9

[Dür06] Dür A.   An improved normalization for the ward reflectance model. *Journal of Graphics, GPU, and Game Tools 11*, 1, pp. 51–59, 2006. DOI: 10.1080/2151237x.2006.10129215. 23

[DVGNK99] Dana K. J., Van Ginneken B., Nayar S. K., and Koenderink J. J.   Reflectance and texture of real-world surfaces. *ACM Transactions on Graphics (TOG) 18*, 1, pp. 1–34, 1999. DOI: 10.1145/300776.300778. 7, 55, 59

[DWd*08] Donner C., Weyrich T., d'Eon E., Ramamoorthi R., and Rusinkiewicz S. A layered, heterogeneous reflectance model for acquiring and rendering human skin. *ACM Transactions on Graphics 27*, 5, pp. 140:1–140:12, December 2008. DOI: 10.1145/1409060.1409093. 9

[DWMG15] Dong Z., Walter B., Marschner S., and Greenberg D. P. Predicting appearance from measured microgeometry of metal surfaces. *ACM Transactions on Graphics 35*, 1, pp. 9:1–9:13, December 2015. DOI: 10.1145/2815618. 50

[DWT*10] Dong Y., Wang J., Tong X., Snyder J., Lan Y., Ben-Ezra M., and Guo B. Manifold bootstrapping for SVBRDF capture. *ACM Transactions on Graphics 29*, 4, pp. 98:1–98:10, July 2010. http://doi.acm.org/10.1145/1778765.1778835 DOI: 10.1145/1778765.1778835. 66

[EBJ*06] Edwards D., Boulos S., Johnson J., Shirley P., Ashikhmin M., Stark M., and Wyman C. The halfway vector disk for BRDF modeling. *ACM Transactions on Graphics 25*, 1, pp. 1–18, January 2006. DOI: 10.1145/1122501.1122502. 27

[EDKM04] Ershov S., Durikovic R., Kolchin K., and Myszkowski K. Reverse engineering approach to appearance-based design of metallic and pearlescent paints. *The Visual Computer 20*, 8–9, pp. 586–600, 2004. DOI: 10.1007/s00371-004-0248-0. 32

[EKM01] Ershov S., Kolchin K., and Myszkowski K. Rendering pearlescent appearance based on paint-composition modelling. In *Computer Graphics Forum*, vol. 20, pp. 227–238, Wiley Online Library, 2001. DOI: 10.1111/1467-8659.00515. 31

[Fai05] Fairchild M. J. *Color Appearance Models*, John Wiley & Sons, 2005. DOI: 10.1002/9781118653128. 50

[FCM*08] Francken Y., Cuypers T., Mertens T., Gielis J., and Bekaert P. High quality mesostructure acquisition using specularities. In *CVPR*, 2008. DOI: 10.1109/cvpr.2008.4587782. 64

[FHV15] Filip J., Havlívek M., and Vávra R. Adaptive highlights stencils for modeling of multi-axial BRDF anisotropy. *The Visual Computer*, pp. 1–11, 2015. DOI: 10.1007/s00371-015-1148-1. 38, 49

[Foo97] Foo S. C. *A Gonioreflectometer for Measuring the Bidirectional Reflectance of Material for Use in Illumination Computation*, Ph.D. thesis, Cornell University, 1997. 51

[Fou95] Fournier A. Separating reflection functions for linear radiosity. In *Rendering Techniques 95*, pp. 296–305, Eurographics, Springer Vienna, 1995. DOI: 10.1007/978-3-7091-9430-0_28. 39

[FPBP09] Forés A., Pattanaik S. N., Bosch C., and Pueyo X. BRDFlab: A general system for designing BRDFs. In *19th Congreso Español de Informática Gráfica (CEIG09)*, San Sebastián, Spain, 2009. DOI: 10.2312/LocalChapterEvents/CEIG/CEIG09/153-160. 45, 46

[FV14] Filip J. and Vávra R. Template-based sampling of anisotropic BRDFs. *Computer Graphics Forum (Proceedings of Pacific Graphics)*, 2014. DOI: 10.1111/cgf.12477. 45

[FVK14] Filip J., Vávra R., and Krupička M. Rapid material appearance acquisition using consumer hardware. *Sensors 14*, 10, pp. 19785–19805, 2014. DOI: 10.3390/s141019785. 7

[GAHO07] Ghosh A., Achutha S., Heidrich W., and O'Toole M. BRDF acquisition with basis illumination. In *Computer Vision, (ICCV). IEEE 11th International Conference on*, pp. 1–8, October 2007. http://dx.doi.org/10.1109/ICCV.2007.4408935 DOI: 10.1109/iccv.2007.4408935. 57, 60

[GBS14] Guarnera G. C., Bianco S., and Schettini R. Absolute colorimetric characterization of a DSLR camera. In *Proc. SPIE*, vol. 9023, pp. 90230U–90230U–7, 2014. DOI: 10.1117/12.2042172. 50

[GCHS05] Goldman D. B., Curless B., Hertzmann A., and Seitz S. M. Shape and spatially-varying BRDFs from photometric stereo. In *Proc. of the 10th IEEE International Conference on Computer Vision (ICCV) Volume 1*, pp. 341–348, IEEE Computer Society, Washington, DC, 2005. DOI: 10.1109/iccv.2005.219. 49

[GCP*09]   Ghosh A., Chen T., Peers P., Wilson C. A., and Debevec P. Estimating specular roughness and anisotropy from second order spherical gradient illumination. In *Proc. of the 20th Eurographics Conference on Rendering (EGSR)*, pp. 1161–1170, Eurographics Association, Aire-la-Ville, Switzerland, 2009. DOI: 10.1111/j.1467-8659.2009.01493.x. 59, 62, 65

[GCP*10]   Ghosh A., Chen T., Peers P., Wilson C. A., and Debevec P. Circularly polarized spherical illumination reflectometry. *ACM Transactions on Graphics 29*, 6, pp. 162:1–162:12, December 2010. DOI: 10.1145/1882261.1866163. 62

[Geo03]   Georghiades A. S. Recovering 3D shape and reflectance from a small number of photographs. In *Proc. of the 14th Eurographics Workshop on Rendering (EGRW)*, pp. 230–240, Eurographics Association, Aire-la-Ville, Switzerland, 2003. DOI: 10.2312/EGWR/EGWR03/230-240. 49

[GHAO10]   Ghosh A., Heidrich W., Achutha S., and O'Toole M. A basis illumination approach to BRDF measurement. *International Journal of Computer Vision 90*, 2, pp. 183–197, 2010. DOI: 10.1007/s11263-008-0151-7. 57, 60

[GHP*08]   Ghosh A., Hawkins T., Peers P., Frederiksen S., and Debevec P. Practical modeling and acquisition of layered facial reflectance. In *ACM Transactions on Graphics (TOG)*, vol. 27, p. 139, 2008. DOI: 10.1145/1409060.1409092. 28

[GLL*04]   Goesele M., Lensch H. P. A., Lang J., Fuchs C., and Seidel H.-P. Disco: Acquisition of translucent objects. *ACM Transactions on Graphics 23*, 3, pp. 835–844, August 2004. DOI: 10.1145/1015706.1015807. 8

[GLS04]   Goesele M., Lensch H., and Seidel H.-P. Validation of color managed 3D appearance acquisition. In *Color and Imaging Conference*, vol. 2004, pp. 265–270, Society for Imaging Science and Technology, 2004. 49

[GMD10]   Geisler-Moroder D. and Dür A. A new ward BRDF model with bounded albedo. *Computer Graphics Forum 29*, 4, pp. 1391–1398, 2010. DOI: 10.1111/j.1467-8659.2010.01735.x. 23, 25

[GPDG12]   Guarnera G., Peers P., Debevec P., and Ghosh A. Estimating surface normals from spherical stokes reflectance fields. In *Computer Vision, (ECCV). Workshops and Demonstrations*, Fusiello A., Murino V., Cucchiara R., (Eds.), vol. 7584 of *Lecture Notes in Computer Science*, pp. 340–349, Springer Berlin Heidelberg, 2012. DOI: 10.1007/978-3-642-33885-4. 62

[Gla94]   Glassner, Andrew S. *Principles of Digital Image Synthesis*, Vol II, Morgan Kaufmann, 1994. DOI: 10.1016/c2009-0-27624-0. 3

[GTGB84] Goral C. M., Torrance K. E., Greenberg D. P., and Battaile B. Modeling the interaction of light between diffuse surfaces. *SIGGRAPH Computer Graphics 18*, 3, pp. 213–222, January 1984. DOI: 10.1145/964965.808601. 13

[GTHD03] Gardner A., Tchou C., Hawkins T., and Debevec P. Linear light source reflectometry. *ACM Transactions on Graphics 22*, 3, pp. 749–758, July 2003. `http://doi.acm.org/10.1145/882262.882342` DOI: 10.1145/882262.882342. 63, 64

[GHCG17] Guarnera G. C., Hall P., Chesnais A., and Glencross M. Woven fabric model creation from a single image. *ACM Transactions on Graphics 36*, 5, pp. 165:1–165:13, October 2017. DOI: 10.1145/3132187. 16

[Hai91] Haines E. A. Beams o' light: Confessions of a hacker. *SIGGRAPH Course Notes—Frontiers in Rendering*, July 1991. 16

[HBV03] Hao X., Baby T., and Varshney A. Interactive subsurface scattering for translucent meshes. In *Proc. of the Symposium on Interactive 3D Graphics (I3D)*, pp. 75–82, ACM, New York, 2003. DOI: 10.1145/641480.641497. 9

[HDCD15] Heitz E., Dupuy J., Crassin C., and Dachsbacher C. The SGGX microflake distribution. *ACM Transactions on Graphics 34*, 4, pp. 48:1–48:11, July 2015. DOI: 10.1145/2766988. 33

[Hei14] Heitz E. Understanding the masking-shadowing function in microfacet-based BRDFs. *Journal of Computer Graphics Techniques (JCGT) 3*, 2, pp. 48–107, June 2014. 15

[HF13] Haindl M. and Filip J. Spatially varying bidirectional reflectance distribution functions. In *Visual Texture*, pp. 119–145, Advances in Computer Vision and Pattern Recognition, Springer London, 2013. DOI: 10.1007/978-1-4471-4902-6_6. 7, 8

[HK93] Hanrahan P. and Krueger W. Reflection from layered surfaces due to subsurface scattering. In *Proc. of the 20th Annual Conference on Computer Graphics and Interactive Techniques (SIGGRAPH)*, pp. 165–174, ACM, New York, 1993. DOI: 10.1145/166117.166139. 50

[HLHZ08] Holroyd M., Lawrence J., Humphreys G., and Zickler T. A photometric approach for estimating normals and tangents. *ACM Transactions on Graphics (Proceedings of SIGGRAPH Asia) 27*, 5, 2008. DOI: 10.1145/1457515.1409086. 49

[HM12] Haindl M. and Filip J. V. R. Digital material appearance: The curse of tera-bytes. *ERCIM News*, 90, pp. 49–50, 2012. 45

[HP03] Han J. Y. and Perlin K. Measuring bidirectional texture reflectance with a kalei-doscope. *ACM Transactions on Graphics 22*, 3, pp. 741–748, July 2003. DOI: 10.1145/882262.882341. 7, 49, 67

[HR76] Hsia J. J. and Richmond J. C. Bidirectional reflectometry. Part i: A high res-olution laser bidirectional reflectometer with results on several optical coatings. *Journal of Reseach of the National Bureau of Standards-A. Physics and Chemistry A 80*, pp. 189–205, 1976. DOI: 10.6028/jres.080a.021. 51

[HS05] Hertzmann A. and Seitz S. M. Example-based photometric stereo: Shape reconstruction with general, varying BRDFs. *Pattern Analysis and Ma-chine Intelligence, IEEE Transactions on 27*, 8, pp. 1254–1264, 2005. DOI: 10.1109/tpami.2005.158. 49

[HTSG91] He X. D., Torrance K. E., Sillion F. X., and Greenberg D. P. A comprehensive physical model for light reflection. *SIGGRAPH Computer Graphics 25*, 4, pp. 175–186, July 1991. DOI: 10.1145/127719.122738. 30, 51

[JAM*10] Jakob W., Arbree A., Moon J. T., Bala K., and Marschner S. A radiative transfer framework for rendering materials with anisotropic structure. *ACM Transactions on Graphics 29*, 4, pp. 53:1–53:13, July 2010. DOI: 10.1145/1778765.1778790. 16

[JHY*14] Jakob W., Hašan M., Yan L.-Q., Lawrence J., Ramamoorthi R., and Marschner S. Discrete stochastic microfacet models. *ACM Transactions on Graphics 33*, 4, pp. 115:1–115:10, July 2014. http://dx.doi.org/10.1145/2601097.2601186 DOI: 10.1145/2601097.2601186. 36

[JMLH01] Jensen H. W., Marschner S. R., Levoy M., and Hanrahan P. A practical model for subsurface light transport. In *Proc. of the 28th Annual Conference on Com-puter Graphics and Interactive Techniques (SIGGRAPH)*, pp. 511–518, ACM, New York, 2001. DOI: 10.1145/383259.383319. 8, 9

[KK89] Kajiya J. T. and Kay T. L. Rendering fur with three dimensional tex-tures. *SIGGRAPH Computer Graphics 23*, 3, pp. 271–280, July 1989. DOI: 10.1145/74334.74361. 16

[KK08] Kim M. H. and Kautz J. Characterization for high dynamic range imag-ing. *Computer Graphics Forum 27*, 2, pp. 691–697, 2008. DOI: 10.1111/j.1467-8659.2008.01167.x. 50

[KM99] Kautz J. and McCool M. D. Interactive rendering with arbitrary BRDFs using separable approximations. In *Rendering Techniques 99*, pp. 247–260, Springer, 1999. DOI: 10.1007/978-3-7091-6809-7_22. 41

[SAA15] Ström, J., Åström, K., and Akenine-Möller, T. Immersive linear algebra. http://http://immersivemath.com/ila/index.html, 2017. 3

[KRP*15] Klehm O., Rousselle F., Papas M., Bradley D., Hery C., Bickel B., Jarosz W., and Beeler T. Recent advances in facial appearance capture. *Computer Graphics Forum (Proceedings of Eurographics) 34*, 2, pp. 709–733, May 2015. DOI: 10.1111/cgf.12594. 9

[KRS*15] Kettner L., Raab M., Seibert D., Jordan J., and Keller A., *The Material Definition Language*, in: R. Klein, H. Rushmeier (Eds.), *Workshop on Material Appearance Modeling*, The Eurographics Association, 2015. http://dx.doi.org/10.2312/mam.20151195 DOI: 10.2312/mam.20151195.

[KSKK10] Kurt M., Szirmay-Kalos L., and Křivánek J. An anisotropic BRDF model for fitting and Monte Carlo rendering. *SIGGRAPH Computer Graphics 44*, 1, pp. 3:1–3:15, February 2010. DOI: 10.1145/1722991.1722996. 34

[KSZ*15] Khungurn P., Schroeder D., Zhao S., Bala K., and Marschner S. Matching real fabrics with micro-appearance models. *ACM Transactions on Graphics 35*, 1, pp. 1:1–1:26, December 2015. DOI: 10.1145/2818648. 16

[LBAD*06] Lawrence J., Ben-Artzi A., DeCoro C., Matusik W., Pfister H., Ramamoorthi R., and Rusinkiewicz S. Inverse shade trees for non-parametric material representation and editing. In *ACM SIGGRAPH Papers*, pp. 735–745, New York, 2006. DOI: 10.1145/1179352.1141949. 43

[Lew94] Lewis R. R. Making shaders more physically plausible. *Computer Graphics Forum 13*, 2, pp. 109–120, 1994. DOI: 10.1111/1467-8659.1320109. 21

[LF97] Lalonde P. and Fournier A. A wavelet representation of reflectance functions. *IEEE Transactions on Visualization and Computer Graphics 3*, 4, pp. 329–336, October 1997. DOI: 10.1109/2945.646236. 38

[LFTG97] Lafortune E. P. F., Foo S.-C., Torrance K. E., and Greenberg D. P. Nonlinear approximation of reflectance functions. In *Proc. of the 24th Annual Conference on Computer Graphics and Interactive Techniques (SIGGRAPH)*, pp. 117–126, ACM, Press/Addison-Wesley Publishing Co., New York, 1997. DOI: 10.1145/258734.258801. 24, 43

[LFTW06] Li H., Foo S.-C., Torrance K. E., and Westin S. H. Automated three-axis gonioreflectometer for computer graphics applications. *Optical Engineering 45*, 4, pp. 043605–043605, 2006. DOI: 10.1117/1.2192787. 51

[LKG*03] Lensch H. P. A., Kautz J., Goesele M., Heidrich W., and Seidel H.-P. Image-based reconstruction of spatial appearance and geometric detail. *ACM Transactions on Graphics 22*, 2, pp. 234–257, April 2003. DOI: 10.1145/636886.636891. 66

[LKYU12] Löw J., Kronander J., Ynnerman A., and Unger J. BRDF models for accurate and efficient rendering of glossy surfaces. *ACM Transactions on Graphics (TOG) 31*, 1, p. 9, 2012. DOI: 10.1145/2077341.2077350. 35, 36, 37

[LRR04] Lawrence J., Rusinkiewicz S., and Ramamoorthi R. Efficient BRDF importance sampling using a factored representation. In *ACM Transactions on Graphics (TOG)*, vol. 23, pp. 496–505, 2004. DOI: 10.1145/1015706.1015751. 17, 42

[LT05] Li H. and Torrance K. E. A practical, comprehensive light reflection model. Tech. pep. PCG-05-03, 2005. 52

[LW94] Lafortune E. P. and Willems Y. D. Using the modified phong reflectance model for physically based rendering. Tech. rep., 1994. 25

[MAA01] McCool M. D., Ang J., and Ahmad A. Homomorphic factorization of BRDFs for high-performance rendering. In *Proc. of the 28th Annual Conference on Computer Graphics and Interactive Techniques (SIGGRAPH)*, pp. 171–178, ACM, New York, 2001. DOI: 10.1145/383259.383276. 42

[Mat03] Matusik W. *A Data-driven Reflectance Model*. Ph.D. thesis, Massachusetts Institute of Technology, 2003. DOI: 10.1145/882262.882343. 45

[MD98] Marschner S. R. Inverse rendering for computer graphics. Tech. rep., 1998. 53

[MGW01] Malzbender T., Gelb D., and Wolters H. Polynomial texture maps. In *Proc. of the 28th Annual Conference on Computer Graphics and Interactive Techniques (SIGGRAPH)*, pp. 519–528, ACM, New York, 2001. DOI: 10.1145/383259.383320. 57, 61

[MHH*12] McAuley S., Hill S., Hoffman N., Gotanda Y., Smits B., Burley B., and Martinez A. Practical physically-based shading in film and game production. In *ACM SIGGRAPH Courses*, pp. 10:1–10:7, New York, 2012. DOI: 10.1145/2343483.2343493. 33, 45, 46

[MHP*07] Ma W.-C., Hawkins T., Peers P., Chabert C.-F., Weiss M., and Debevec P. Rapid acquisition of specular and diffuse normal maps from polarized spherical gradient illumination. In *Proc. of the 18th Eurographics Conference on Rendering Techniques (EGSR)*, pp. 183–194, Eurographics Association, Aire-la-Ville, Switzerland, 2007. DOI: 10.2312/EGWR/EGSR07/183-194. 59

[MPBM03a] Matusik W., Pfister H., Brand M., and McMillan L. A data-driven reflectance model. *ACM Transactions on Graphics 22*, 3, pp. 759–769, July 2003. DOI: 10.1145/882262.882343. 23, 33, 39, 53, 55, 56

[MPBM03b] Matusik W., Pfister H., Brand M., and McMillan L. Efficient isotropic BRDF measurement. In *Proc. of the 14th Eurographics Workshop on Rendering (EGRW)*, pp. 241–247, Eurographics Association, Aire-la-Ville, Switzerland, 2003. 40

[MSY07] Mukaigawa Y., Sumino K., and Yagi Y. Multiplexed illumination for measuring BRDF using an ellipsoidal mirror and a projector. In *Computer Vision (ACCV)*, pp. 246–257, Springer, 2007. DOI: 10.1007/978-3-540-76390-1_25. 56, 57, 60

[MWL*99] Marschner S. R., Westin S. H., Lafortune E. P., Torrance K. E., and Greenberg D. P. Image-based BRDF measurement including human skin. In *Rendering Techniques*, pp. 131–144, Springer, 1999. DOI: 10.1007/978-3-7091-6809-7_13. 53, 54

[NDM05] Ngan A., Durand F., and Matusik W. Experimental analysis of BRDF models. In *Proc. of the 16th Eurographics Conference on Rendering Techniques (EGSR)*, pp. 117–126, Eurographics Association, Aire-la-Ville, Switzerland, 2005. http://dx.doi.org/10.2312/EGWR/EGSR05/117-126 31, 53, 57

[NL11] Nishino K. and Lombardi S. Directional statistics-based reflectance model for isotropic bidirectional reflectance distribution functions. *Journal of the Optical Society of America A 28*, 1, pp. 8–18, January 2011. DOI: 10.1364/josaa.28.000008. 22

[NNSK99] Neumann L., Neumannn A., and Szirmay-Kalos L. Compact metallic reflectance models. *Computer Graphics Forum 18*, 3, pp. 161–172, 1999. DOI: 10.1111/1467-8659.00337. 23, 25

[NRH*77] Nicodemus F., Richmond J., Hsia J., Ginsberg I., and Limperis T. Geometrical considerations and nomenclature for reflectance. *Natl. Bur. Stand. Rep., NBS MN-160*, 1977. DOI: 10.6028/nbs.mono.160. 8

[NZV*11] Naik N., Zhao S., Velten A., Raskar R., and Bala K. Single view reflectance capture using multiplexed scattering and time-of-flight imaging. *ACM Transactions on Graphics 30*, 6, pp. 171:1–171:10, December 2011. http://doi.acm.org/10.1145/2070781.2024205 DOI: 10.1145/2070781.2024205. 54

[OKBG08] Ozturk A., Kurt M., Bilgili A., and Gungor C. Linear approximation of bidirectional reflectance distribution functions. *Computers and Graphics 32*, 2, pp. 149–158, 2008. DOI: 10.1016/j.cag.2008.01.004. 26, 43, 53

[ON94] Oren M. and Nayar S. K. Generalization of lambert's reflectance model. In *Proc. of the 21st Annual Conference on Computer Graphics and Interactive Techniques*, pp. 239–246, ACM, 1994. DOI: 10.1145/192161.192213. 31

[PFTV88] Press W. H., Flannery B. P., Teukolsky S. A., and Vetterling W. T. *Numerical Recipes in C: The Art of Scientific Computing*, Cambridge University Press, New York, 1988. 39

[PH89] Perlin K. and Hoffert E. M. Hypertexture. *SIGGRAPH Computer Graphics 23*, 3, pp. 253–262, July 1989. DOI: 10.1145/74334.74359. 16

[Pho75] Phong B. T. Illumination for computer generated pictures. *Communication ACM 18*, 6, pp. 311–317, June 1975. DOI: 10.1145/360825.360839. 21, 25

[PSCS*12] Pacanowski R., Salazar Celis O., Schlick C., Granier X., Poulin P., and Cuyt A. Rational BRDF. *Visualization and Computer Graphics, IEEE Transactions on 18*, 11, pp. 1824–1835, November 2012. DOI: 10.1109/tvcg.2012.73. 44

[RCH12] Riviere N., Ceolato R., and Hespel L. Multispectral polarized BRDF: Design of a highly resolved reflectometer and development of a data inversion method. *Opt. Appl. 42*, 2012. DOI: 10.5277/oa120101. 52

[RLCP] Rushmeier H., Lockerman Y. D., Cartwright L., and Pitera D. Experiments with a low-cost system for computer graphics material model acquisition. DOI: 10.1117/12.2082895. 49

[RMS*08] Rump M., Müller G., Sarlette R., Koch D., and Klein R. Photo-realistic rendering of metallic car paint from image-based measurements. *Computer Graphics Forum 27*, 2, pp. 527–536, April 2008. DOI: 10.1111/j.1467-8659.2008.01150.x. 33, 59

[RPG15] Riviere J., Peers P., and Ghosh A. Mobile surface reflectometry. *Computer Graphics Forum*, pp. n/a–n/a, 2015. http://dx.doi.org/10.1111/cgf.12719 DOI: 10.1111/cgf.12719. 8, 65, 66

[Rus98] Rusinkiewicz S. A new change of variables for efficient BRDF representation. In *Rendering Techniques 98*, Drettakis G., Max N., (Eds.), Eurographics Book Series. Springer-Verlag, Wien, Austria, 1998. *Proc. of the Workshop*, Vienna, Austria, June 29–July 1, 1998. DOI: 10.1007/978-3-7091-6453-2. 10, 38, 39, 40, 44

[RVZ08] Romeiro F., Vasilyev Y., and Zickler T. Passive reflectometry. In *Proc. of the 10th European Conference on Computer Vision: Part IV (ECCV)*, pp. 859–872, Berlin, Heidelberg, Springer-Verlag, 2008. DOI: 10.1007/978-3-540-88693-8_63. 40

[RWS*11] Ren P., Wang J., Snyder J., Tong X., and Guo B. Pocket reflectometry. *ACM Transactions on Graphics 30*, 4, pp. 45:1–45:10, July 2011. DOI: 10.1145/2010324.1964940. 63

[SAS05] Stark M., Arvo J., and Smits B. Barycentric parameterizations for isotropic BRDFs. *Visualization and Computer Graphics, IEEE Transactions on 11*, 2, pp. 126–138, March 2005. DOI: 10.1109/tvcg.2005.26. 10

[Sch94] Schlick C. An inexpensive BRDF model for physically-based rendering. *Computer Graphics Forum 13*, 3, pp. 233–246, 1994. DOI: 10.1111/1467-8659.1330233. 13, 27

[SDSG13] Schregle R., Denk C., Slusallek P., and Glencross M. Grand challenges: Material models in automotive. In *Proc. of the Eurographics Workshop on Material Appearance Modeling: Issues and Acquisition (MAM)*, pp. 1–6, Eurographics Association, Aire-la-Ville, Switzerland, 2013. DOI: 10.2312/MAM.MAM2013.001-006. 2

[Smi67] Smith B. Geometrical shadowing of a random rough surface. *Antennas and Propagation, IEEE Transactions on 15*, 5, pp. 668–671, 1967. DOI: 10.1109/tap.1967.1138991. 32

[WGK14] Sattler M., Sarlette R., and Klein R. Efficient and realistic visualization of cloth. In *Eurographics Symposium on Rendering*, pp. 156–171, Leuven, Belgium, June, 2003, *Proc. of the 14th Eurographics Workshop on Rendering*, pp. 167–177, Eurographics Association. http://cg.cs.uni-bonn.de/en/projects/btfdbb/download/ubo2003/ 45

[Sta99] Stamminger M. *Finite Element Methods for Global Illumination Computations*, Herbert Utz Verlag, 1999. 46

[SZC*07] Sun X., Zhou K., Chen Y., Lin S., Shi J., and Guo B. Interactive relighting with dynamic BRDFs. *ACM Transactions on Graphics 26*, 3, July 2007. DOI: 10.1145/1276377.1276411. 43

[TFG*13] Tunwattanapong B., Fyffe G., Graham P., Busch J., Yu X., Ghosh A., and Debevec P. Acquiring reflectance and shape from continuous spherical harmonic illumination. *ACM Transactions on Graphics 32*, 4, pp. 109:1–109:12, July 2013. http://doi.acm.org/10.1145/2461912.2461944 DOI: 10.1145/2461912.2461944. 62, 63

[TR75] Trowbridge T. S. and Reitz K. P. Average irregularity representation of a rough surface for ray reflection. *Journal of Optical Society of America 65*, 5, pp. 531–536, May 1975. DOI: 10.1364/josa.65.000531. 33

[TS67]   Torrance K. E. and Sparrow E. M. Theory for off-specular reflection from roughened surfaces. *JOSA 57*, 9, pp. 1105–1112, 1967. DOI: 10.1364/josa.57.001105. 28, 51

[Tuc66]   Tucker L. Some mathematical notes on three-mode factor analysis. *Psychometrika 31*, 3, pp. 279–311, 1966. DOI: 10.1007/bf02289464. 43

[Vea97]   Veach E. *Robust Monte Carlo Methods for Light Transport Simulation*. Ph.D. thesis, Stanford University, 1997. 13

[War92]   Ward G. J. Measuring and modeling anisotropic reflection. *SIGGRAPH Computer Graphics 26*, 2, pp. 265–272, July 1992. DOI: 10.1145/142920.134078. 23, 25, 43, 49, 55, 58, 63

[WAT92]   Westin S. H., Arvo J. R., and Torrance K. E. Predicting reflectance functions from complex surfaces. *SIGGRAPH Computer Graphics 26*, 2, pp. 255–264, July 1992. DOI: 10.1145/142920.134075. 38, 50

[WDR11]   Wu H., Dorsey J., and Rushmeier H. A sparse parametric mixture model for BTF compression, editing and rendering. In *Computer Graphics Forum*, vol. 30, pp. 465–473, Wiley Online Library, 2011. DOI: 10.1111/j.1467-8659.2011.01890.x. 8

[WEV02]   Ward G. and Eydelberg-Vileshin E. Picture perfect RGB rendering using spectral prefiltering and sharp color primaries. In *Proc. of the 13th Eurographics Workshop on Rendering (EGRW)*, pp. 117–124, Eurographics Association, Aire-la-Ville, Switzerland, 2002. 50

[WKB12]   Ward G., Kurt M., and Bonneel N. Framework for sharing and rendering real-world bidirectional scattering distribution functions. Tech. Rep. LBNL-5954E, Lawrence Berkeley National Laboratory, October 2012. 45

[WKB14]   Ward G., Kurt M., and Bonneel N. Reducing anisotropic BSDF measurement to common practice. In *Proc. of the Eurographics Workshop on Material Appearance Modeling: Issues and Acquisition (MAM)*, pp. 5–8, Eurographics Association, Aire-la-Ville, Switzerland, 2014. 45

[WGK14]   Weinmann M., Gall J., and Klein R. Material classification based on training data synthesized using a BTF database. In *Computer Vision (ECCV), 13th European Conference*, Zurich, Switzerland, September 6–12, 2014. *Proc., Part III*, pp. 156–171, Springer International Publishing. http://cg.cs.uni-bonn.de/en/projects/btfdbb/download/ubo2014/ DOI: 10.1007/978-3-319-10578-9_11. 45

[WLL*09] Weyrich T., Lawrence J., Lensch H., Rusinkiewicz S., and Zickler T. Principles of appearance acquisition and representation. *Foundations and Trends® in Computer Graphics and Vision 4*, 2, pp. 75–191, 2009. DOI: 10.1145/1401132.1401234. 29

[WLT04] Westin S. H., Li H., and Torrance K. E. A comparison of four BRDF models. In *Eurographics Symposium on Rendering*, pp. 1–10, 2004. 24

[WMLT07] Walter B., Marschner S. R., Li H., and Torrance K. E. Microfacet models for refraction through rough surfaces. In *Proc. of the 18th Eurographics conference on Rendering Techniques*, pp. 195–206, Eurographics Association, 2007. 8, 32, 33, 65

[Woo80] Woodham R. J. Photometric method for determining surface orientation from multiple images. *Optical Engineering 19*, 1, p. 191139, 1980. DOI: 10.1117/12.7972479. 49

[WS82] Wyszecki G. and Stiles W. S. *Color Science: Concepts and Methods, Quantitative Data and Formulae*, vol. 8, 1982. DOI: 10.1063/1.3035025. 49

[WSM11] Wang C.-P., Snavely N., and Marschner S. Estimating dual-scale properties of glossy surfaces from step-edge lighting. In *Proc. of the SIGGRAPH Asia Conference (SA)*, pp. 172:1–172:12, ACM, New York, 2011. http://doi.acm.org/10.1145/2024156.2024206 DOI: 10.1145/2070752.2024206. 64

[WW07] Weidlich A. and Wilkie A. Arbitrarily layered micro-facet surfaces. In *Proc. of the 5th International Conference on Computer Graphics and Interactive Techniques in Australia and Southeast Asia (GRAPHITE)*, pp. 171–178, ACM, New York, 2007. DOI: 10.1145/1321261.1321292. 37

[XCL*01] Xu Y.-Q., Chen Y., Lin S., Zhong H., Wu E., Guo B., and Shum H.-Y. Photorealistic rendering of knitwear using the lumislice. In *Proc. of the 28th Annual Conference on Computer Graphics and Interactive Techniques (SIGGRAPH)*, pp. 391–398, ACM, New York, 2001. DOI: 10.1145/383259.383303. 16

[ZJMB11] Zhao S., Jakob W., Marschner S., and Bala K. Building volumetric appearance models of fabric using micro CT imaging. *ACM Transactions on Graphics 30*, 4, pp. 44:1–44:10, July 2011. DOI: 10.1145/2010324.1964939. 16

[ZREB06] Zickler T., Ramamoorthi R., Enrique S., and Belhumeur P. N. Reflectance sharing: Predicting appearance from a sparse set of images of a known shape. *IEEE Transactions on Pattern Anal. Mach. Intell. 28*, 8, pp. 1287–1302, August 2006. DOI: 10.1109/tpami.2006.170. 49

Authors' Biographies

DAR'YA GUARNERA

Dar'ya Guarnera is a Research Associate at Norwegian University of Science and Technology. She obtained her Ph.D. in Computer Science at Loughborough University, UK. She obtained her Postgraduate Certificate in Education and first-class degree in Computer Science from Liverpool Hope University. In 2006, she received the Student of the Year award upon graduation from Liverpool Community College in 3D modeling. She also obtained a B.Sc. in Architecture from the Odessa State Academy of Civil Engineering and Architecture. Her interests include 3D Architectural Visualization, virtual materials, digital art, and sculpture. She is a guest lecturer at Liverpool Hope University and part-time lecturer at Liverpool Community College.

GIUSEPPE CLAUDIO GUARNERA

Giuseppe Claudio Guarnera, Ph.D., is a Research Associate at Norwegian University of Science and Technology. His current research interests include appearance modeling and human perception of materials. Previously he was a postdoctoral Research Associate at Loughborough University, focusing on fabric appearance modeling and acquisition. He received his Ph.D. in Computer Science from the University of Catania, Italy, with a doctoral dissertation in the area of Computer Vision and Pattern Recognition, partially achieved while he was a Visiting Researcher at the USC Institute for Creative Technologies under the supervision of Dr. Abhijeet Ghosh and Prof. Paul Debevec.